ORIGINAL PUBLICATION IN FRENCH BY
FRENCH INSTITUTE OF PETROLEUM
AND
EDITIONS TECHNIP

SEISMIC FILTERING

Translated by
Nathan Rothenburg

Edited by
Robert Van Nostrand

Society of Exploration Geophysicists

Originally published in 1966 as Le Filtrage En Sismique
© *1966, Editions Technip — Paris.*

Society of Exploration Geophysicists, Tulsa, Oklahoma 74101.
Translation © *1971 by Society of Exploration Geophysicists.*
All rights reserved. Published 1971
Printed in the United States of America

ALPHABETICAL LIST OF CONTRIBUTORS

J. CASSAND

B. DAMOTTE

A. FONTANEL

G. GRAU

CH. HEMON

M. LAVERGNE

TABLE OF CONTENTS

Chapter 8
SEIMIC EMISSION BY VIBRATORS

SEISMIC EMISSION BY VIBRATORS 198

J. Cassand and M. Lavergne

FOREWORD

A seminar on problems of seismic filtering was given by the French Petroleum Institute during January 11-22, 1965 and October 25 — November 5, 1965. Some twenty-five lectures were delivered at this seminar by engineers of the Geophysical Division who also conducted demonstrations on filtering and discussed the results.

Our purpose is to present here the lectures themselves or enlarged texts on the same subjects. It was not possible to follow a logical sequence: to save time we thought it preferable to incorporate the texts as they became available. Others will appear in later volumes. Apart from this, the general layout of the work will be found on the following page.

It will be noted that the general theme concerns filtering in the broad sense and not merely the different operations applied to seismic sections during processing at playback centers. We therefore construe filtering here to be the whole system of transformations imparted to seismic data by the earth, the recording instruments and the processing. That is why we have studied the characteristics of analog and digital recording equipment which have a bearing on the preservation of information. Likewise, we have considered numerical calculation as a process which band-limits frequencies and causes the signal-to-noise ratio to vary. Finally, we have briefly discussed in the same sense, seismometers and their ground coupling and the emission of vibrators. In wave propagation, we have omitted everything which does not have a direct bearing on seismic resolution, surface waves being considered here only in their gross aspect, without having to study the mechanics of their production in detail. In principle, everything is then directed to the same objective: to bring to the attention of geophysicists the reason for things, so as to impress upon them the inherent limitations of the seismic method and how it can be improved.

It will be useful to think about these different points when digital recording becomes widespread in certain countries while others are converting to analog recording on magnetic tape.

In both cases, it is a question of knowing where we are going, what is feasible in practice and the predictions allowed by theory. Such is the thinking which has prompted this work.

Assignments which appear in certain chapters form an integral part of the text and the reader should make an effort to do them. Their answers are given separately.

The bibliographies offer no pretense at being complete. We have simply mentioned here works or articles which are deemed indispensable to an understanding of the subject.

G. GRAU

Chapter 1

INTRODUCTION TO ONE-DIMENSIONAL SPECTRA

G. GRAU

In this chapter, spectra of functions with a single variable will be studied, for example those that are calculated when seeking the frequency content of an individual seismic trace. These few pages should be regarded only as an introduction to spectra. The special properties of spectra encountered in seismic work will be described in the chapter on signal and noise.

Furthermore, the theorems which are given in the present introduction will be supplemented by others which may be stated only after spectral distributions and convolution are described. At this time, a more systematic account of spectral properties will be given.

PHYSICAL INTRODUCTION TO THE SPECTRAL CONCEPT

For the purpose of introduction, an example will be chosen which will serve us later in the description of noise and in the explanation of seismic filtering. The concepts of frequency and the representation of periodic phenomena by systems of rotating vectors will be assumed to be known.

Let us suppose that a large number of very closely spaced seismometers S are located on a straight line (L), in order to study the waves generated by a sufficiently distant source with constant characteristics. The source may be an eccentric rotating device M, which excites the ground in sinusoidal fashion at each turn (Figure 1). Let us first assume that M is located at a very large distance from the profile (L) and that propagation takes place without distortion, the ground surface being flat and the subsurface homogeneous and isotropic. This case is similar to that in radio-astronomy, M being a source in space and S the

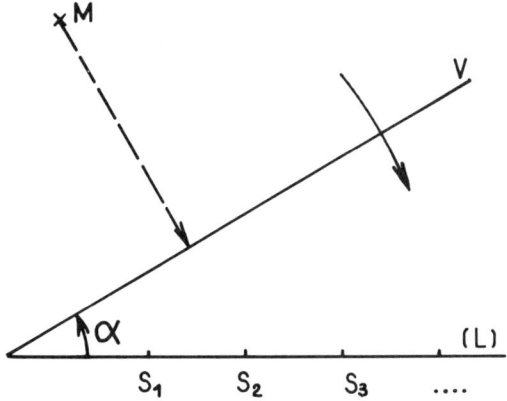

Figure 1.   An assumed plane wavefront from a distant source M striking a line L of seismometers at an angle α.

1

receiver of hertzian waves; it is also analogous to conventional astronomy with M as a bright star and S as one of many photo-sensitive cells (Jennison, 1961). We can even imagine that M generates surface waves in a large expanse of water, the ideal case being that of a well-defined swell (M is then at infinity) observed from the line (L) far from the shore in a deep sea without wind or currents. In a similar situation, the waves received at (L) are practically plane and their wavefronts form a constant angle α with (L). A given wavefront that arrives at (L) with the velocity V, which is a function of the medium, will traverse the line of seismometers (L), with the apparent velocity $V_\alpha = V/\sin \alpha$. Since the excitation produced by the source is sinusoidal, we write it as $u_\alpha e^{j\omega t}$, $\omega = 2\pi f$ being the angular frequency. If the movement at S_1 is $u_\alpha e^{j(\omega t - \varphi)}$ the movement at S_2 is

$$u_\alpha e^{j(\omega t - \varphi - \omega \overline{S_1 S_2}/V_\alpha)}$$

and likewise for all the seismometers of the line (L). If we set $V_\alpha/f = \lambda_\alpha$, the apparent wavelength along (L), we see that at a given instant the movement is determined by the function:

$$u_\alpha e^{-j\omega x/V_\alpha} = u_\alpha e^{-2\pi j x/\lambda_\alpha},$$

x being the distance on (L) measured from an origin, S_1 for example.

For each angle α, hence for each value of λ_α, there is a corresponding function $u_\alpha e^{-2\pi j x/\lambda_\alpha}$ which describes the relative phase of the movement recorded by the different seismometers. It can be seen that the smaller α becomes, the slower the phase $-2\pi x/\lambda_\alpha$ changes with varying x; and conversely, as α increases, the faster the phase changes. The amplitude of the movement itself remains constant and equal to u_α.

If there are now several sources, since the propagation and reception are assumed to be linear, the function which describes the movement on (L) is the sum of the functions corresponding to each of the sources. For example, if two sources with amplitudes u_α and u_β produce wavefronts arriving at (L) with angles α and β, the function will be:

$$u_\alpha e^{-2\pi j x/\lambda_\alpha} + u_\beta e^{-2\pi j x/\lambda_\beta}.$$

In this case, it is seen that the phase is no longer proportional to x, and the amplitude measured along (L) no longer remains constant.

It is simple to generalize in the case of a continuous distribution of sources at great distance, whose amplitudes and phases will be given by a function g of the angle α which may be real or complex. The movement on (L) is defined by:

$$G(x) = \int_{-\pi/2 + \alpha_0}^{\pi/2 - \alpha_0} g(\alpha) e^{-2\pi j x/\lambda_\alpha} d\alpha.$$

The term α_0 (Figure 2) is intended to locate the sources at a large distance from all the points of (L), in keeping with the assumptions. It will be noticed that Jennison (1961, p. 15) takes $-2\pi jx\alpha$, with our notation, as the argument of the complex exponential and not $-2\pi jx/\lambda_\alpha$. Thus the implicit assumption is made that α is

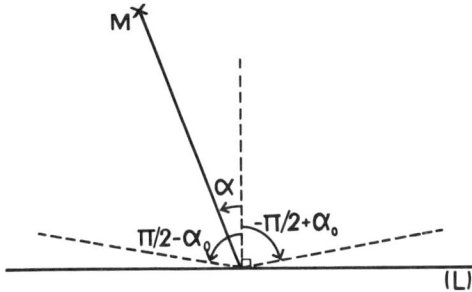

Figure 2.   Limits of integration over a continuously distributed source at great distance from the line L.

close to 0 and that x is measured in units numerically equal to V/f in the system of units chosen. The formula given above has the advantage of being homogenous. It is convenient to replace α as the variable, by $k = 1/\lambda_\alpha = (f\sin\alpha)/V$ which will be called the "wavenumber." Function $g(k)$ being a different function from $g(\alpha)$, we then have

$$G(x) = \int_{-k_0}^{k_0} g(k)\, e^{-2\pi jkx}\, dk.$$

Mathematicians will recognize in this integral a form which becomes the Fourier transform of g, if k_0 is made to approach infinity. In the present case, values greater than $(f\cos\alpha_0)/V$ cannot be assigned to k.

Later, we shall study certain properties of the Fourier transform. For the present, we shall content ourselves with considering the problem physically and remember that a function g on (L) corresponds to a unique amplitude G distribution G will be termed the spectrum of g for reasons which will become apparent later.

Here, now, is the inverse problem. Sources are distributed on (L) and their amplitude and phase are described by a function $G(x)$. By virtue of the reciprocity principle, the vibrations emitted by these sources are compounded at infinity to restore a state of motion $g(k)$, it being understood that the angles α are now measured in terms of corresponding k's.

If the phases are sinusoidally distributed along (L) with the wavelength λ, the amplitude of the sources being constant, $g(k)$ reduces to a single nonzero amplitude corresponding to an angle α such that $\sin\alpha = V/(f\lambda)$.

Let us suppose that the line (L) is overlain with sources vibrating with the same amplitude and in-phase over its entire length which is infinite. Motion in a direction perpendicular to (L) and only in this direction will be observed at great distance from (L). If now we imagine that all the sources situated between $-\infty$ and 0 on the abscissa are terminated, $g(k)$ will stretch over the entire extent of the horizon. We shall see later what the amplitude and phase distribution become. This case corresponds to diffraction at infinity of a plane wave by a semi-infinite and perfectly absorbent screen.

In a general way, the spectra $G(x)$, to which reference is made in this physical introduction, are the exact equivalent of diffractions at infinity or Fraunhofer diffractions that are observed in optics. It should, however, be noted in this regard that in optics we do not observe amplitudes, but intensities which are proportional to the square of the light vibration amplitudes, the phases being lost.

We shall be able to experiment with single-dimension, optical diffraction by simulating a uniform and rectilinear source located at infinity with a point source and a cylindrical lens or with a collimated beam from a point source observed through a narrow slit. By placing different one-dimensional objects in the light beam and adjusting to infinity, diffraction figures can be seen. These are spectra of the objects studied, or more precisely, the square of the modulus of these spectra.

SPECTRA AND FOURIER TRANSFORMATION

After this physical introduction we are going to perform a little mathematics, leaving out a certain number of ticklish problems which arise concerning the existence of integrals that will be used. The first part of Arsac (1961) may be consulted on these theoretical questions.

It is shown that, provided certain conditions of regularity are fulfilled, a function $g(t)$ can be written in the form:

$$\int_{-\infty}^{\infty} G(f)\, e^{2\pi jft}\, df, \text{ where } j = \sqrt{-1}.$$

The function $G(f)$ is in general complex and can be written:

$$\int_{-\infty}^{\infty} g(t)\, e^{-2\pi jft}\, dt.$$

This is written in abbreviated form $g(t) \longleftrightarrow G(f)$, which should be read: $g(t)$ has the Fourier transform $G(f)$. The elegant symmetry of the Fourier formulas corresponds to the reciprocity principle of waves in the example of the physical introduction.

Generally, capital letters are reserved for Fourier transforms of quantities directly accessible to experiment, such as functions of time or a space coordinate, which are designated by small letters.

Since we have seen that the Fourier transform of mathematicians corresponds to the spectrum of physicists and enables us to calculate it, we shall say by analogy that if $g \longleftrightarrow G$, G is the spectrum of g.

It is obvious that the two variables t and f are arbitrary. If t is the time, f is the frequency. This is the notation that we shall adopt. In the theory of wavenumber filtering, it will be seen that interesting equations are obtained if time is replaced by distance, then f is replaced by the wavenumber (refer to the physical introduction: the wavenumber $1/\lambda_\alpha$ is the equivalent of f). When t is expressed in seconds, f is expressed in cycles per second or hertz. Likewise, if the distance is reckoned in kilometers, the wavenumber will be measured in cycles per kilometer.

In the physical introduction we have seen that for an isolated source located in a certain direction at angle α with the perpendicular to a straight line (L), there is a corresponding spectrum over this line, proportional to

$$e^{-2\pi jkx},$$

k being the wavenumber associated with azimuth α. The equivalent for the time-frequency pair is the case of a single frequency or a spectral line, as the spectroscopists say. The time function is then:

$$e^{-j\omega t}.$$

If there are several spectral lines, or a continuous distribution of amplitudes as a function of frequency, the function $g(t)$ is the sum of a finite or infinite number of complex exponential functions of the type just shown. To obtain a function of t in the form of $\cos \omega t$, two identical lines with opposite frequencies will be required. Later we shall see how to think of the negative frequency concept.

The actual form of the Fourier integral is

$$G(f) = \int_{-\infty}^{\infty} g(t)\, e^{-j\omega t}\, dt, \quad (\omega = 2\pi f),$$

Therefore, it is easily seen that

$$\boxed{g \longleftrightarrow G, \int_{-\infty}^{\infty} g(t)\, dt = G(0)}\ ;$$

and that conversely,

$$\int_{-\infty}^{\infty} G(f)\, df = g(0)$$.

It follows that if $g \geq 0$, $G(0)$ is the maximum of $G(f)$; if G is real and ≥ 0, $g(0)$ is the maximum of $g(t)$.

Problem 1

Show that if $G(f)$ is the spectrum of $g(t)$, $g(-t)$ is the spectrum of $G(f)$.

The spectrum $G(f)$ is usually a complex function of f that can be written:

$$M(f)\, e^{j\varphi(f)}.$$

The quantity $M(f)$ is termed the modulus and the quantity $\varphi(f)$ is the phase of $G(f)$. One corresponds to the absolute value and the other to the argument of the vector function of f, which represents $G(f)$ in the complex plane. This implies that various equivalent graphical displays can be found for the spectrum. Separate presentations of M and φ as a function of $\omega = 2\pi f$ or of the frequency f are most often employed. We can also represent $Me^{j\varphi}$ by a polar diagram and scale the curve in frequency. We seldom employ displays of the real part $M\cos\varphi$ and imaginary part $M\sin\varphi$ of $G(f)$.

It is important to know that $\varphi(f)$ is, like $M(f)$, indispensable for a complete definition of the function $g(t)$. This can be reviewed in an example that will be treated as an exercise.

Problem 2

Find the modulus M and the phase φ of the spectrum of a square pulse g(t) defined by:

$$\begin{bmatrix} g(t) = 0 & \begin{cases} t < 0, \\ t > T, \end{cases} \\ g(t) = 1 & 0 < t < T. \end{bmatrix}$$

These values are easily substituted into the integral which defines the spectrum $G(f)$ of g(t).

Starting with the modulus M(f), try to synthesize the square pulse; that is, calculate

$$\int_{-\infty}^{\infty} M(f)\, e^{j\omega t}\, df.$$

Problem 3

Calculate the spectrum of the triangular function:

$$\begin{cases} s(t) = 0 & \begin{cases} t < -T, \\ t > T, \end{cases} \\ s(t) = t + T & -T < t < 0, \\ s(t) = T - t & 0 < t < T. \end{cases}$$

As a further illustration, an example is given of a signal s, which could represent a seismic pulse, and two signals s' and s'', one of which has the same modulus as s and zero phase, and the other with the same phase as s but a constant modulus (Figure 3).

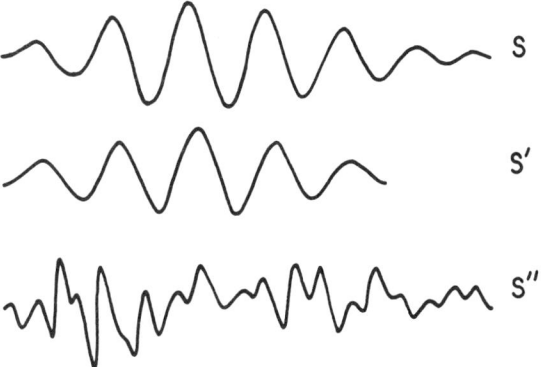

Figure 3. Seismic signal,s together with synthetic signals s' having the same amplitude spectrum and constant phase and s'' having a flat amplitude spectrum and the same phase as s.

Spectrally, if $s \longleftrightarrow S, s' \longleftrightarrow S'$, and $s'' \longleftrightarrow S''$ and if $S = M e^{j\varphi}$, $S' = M$, and $S'' = e^{j\varphi}$. By computing the transforms of M and $e^{j\varphi}$, s' and s'' are obtained. This last signal s'' is very long and only its beginning is shown.

We can observe that s is identical neither to s' nor to s'', which shows that neither the modulus nor the phase is sufficient to define correctly its spectrum.

Nevertheless, it must be noted that the signals which have the same modulus bear a striking similarity, at least when their spectrum affects only a narrow band. If this band were very narrow, the signals would be sinusoidal and would have the same frequency. They would then look exactly alike. It is seen that the resemblance may persist as the bandwidth is broadened progressively.

Later, other results on phase spectra will be seen. Let us continue our investigation of spectral properties.

Problem 4

Show that the real part of the Fourier transform of a real function is even in ω and that the imaginary part is odd in ω.

Problem 5

     Show that the spectrum of a real function has an even modulus and an odd phase in f.

Problem 6

     Show that if g(t) *is real and zero for all values of* t < 0, *it can be calculated from a knowledge of the real part of its spectrum* G(f) *by the integral:*

$$g\,(t) = 2 \int_{-\infty}^{\infty} [\mathrm{Re}\; G\,(f)]\; \cos \omega t\; df = 4 \int_{0}^{\infty} [\mathrm{Re}\; G\,(f)]\; \cos \omega t\; df$$

or from the imaginary part by:

$$g\,(t) = -2 \int_{-\infty}^{\infty} [\mathrm{Im}\; G\,(f)\,]\; \sin \omega t\; df = -4 \int_{0}^{\infty} [\mathrm{Im}\; G\,(f)]\; \sin \omega t\; df.$$

where R = Re G(f) *and* I = Im G(f).

(Write that g(−t) = 0 *for* t > 0 *and make use of the theorem demonstrated in the previous assignment.)*

The spectrum of a function which is zero for negative time can consequently be neither purely real nor purely imaginary. In the light of results from this exercise, one should be able to comment on those of Problem 2.

Problem 7

     Show that if G(f) *is the spectrum of* g(t), *the spectrum of* g(−t) *is* G(−f). *For which case is* G(−f) *equal to* G*(f), *the complex conjugate of* G(f)?

Concerning the last exercise, it may be asked what the meaning of the negative frequencies is. We tend to think intuitively that frequency is always positive since industrial frequency meters have only positive scales. In fact, our intuition is at fault and it is quite justifiable to integrate Fourier integrals from $f = -\infty$ to $f = +\infty$. In the physical introduction on the spectral concept we have seen that the quantity x, being a distance, could take on negative values. There was no difficulty there. When it is no longer a question of distance but of frequency, the mathematics remaining the same, there is no reason not to consider negative frequencies. The physical significance of this is given by the two previous problems. It is known, in fact, that if g is real, there is no need for negative frequencies. The spectrum is not zero for these frequencies, but all the information contained in the spectrum is given by the values taken for the positive frequencies. Furthermore, if t is changed to $-t$, f is changed to $-f$. Besides, in any spectral representation, it may very well be considered that each of the component sinusoids corresponds to a time t and a frequency f, or a time $-t$ and a frequency $-f$. It is in fact the product ωt which serves as the argument for these sinusoids. We therefore cannot dissociate the concept of negative frequency from that of negative time. All this corresponds exactly to the definition of frequency as the number of periods per unit time. The time must really be reckoned negatively in order for the frequency to be negative. It is probably solely for psychological reasons that we are reluctant to imagine negative times and frequencies while readily accepting negative distances. It seems simpler to shift to the right or left of a line marked on the ground, than to take oneself into the past or into the future.

We have seen above that in order to obtain a function $g(t)$ equal to cos $\omega_0 t$, two spectral lines are required, one with frequency $\omega_0/2\pi$, the other with $-\omega_0/2\pi$. This evidently arises from the definition of the complex exponential transform, the arbitrary nature of which may be questioned. The optical diffraction experiment should however reassure us on this point. In fact, if we observe the spectrum of a seismic trace by an optical diffraction experiment, symmetrical patterns with respect to the zero frequency are readily found.

Make a reproduction of any seismic trace of two to three seconds length on transparent film, and by replication of this trace construct a type of seismic section. The dimensions should be such that one second is represented by about two millimeters on the film. Place this microfilm in front of the eye and observe a linear, monochromatic source situated at infinity so that it appears to be parallel to the timing lines. Adjust to infinity. Verify that the spectrum is perfectly symmetrical. Observe the spectrum for a pure sinusoid. Further details on spectra obtained through optical diffraction will be given later.

SOME PHASE CONSIDERATIONS

We have seen above that knowledge of the phase spectrum is just as important as that of its modulus. It is worthwhile reviewing the kind of information contained in the function $\varphi(f)$.

Let us first see what happens when the phase is identically zero; then all the component sinusoids are in-phase for zero time and the time function will have a maximum here. For times different from zero the sinusoids combine in various ways according to their amplitudes.

Problem 8

Find the general form of the phase spectrum for the function g(t) *which is real and symmetrical with respect to the axis* $t = t_0$.

Problem 9

Show mathematically that a shift in the time origin of t_0 *towards positive t corresponds to a multiplication of the spectrum by* $e^{j\omega t_0}$. *Employ this property for a new explanation of the previous problem.*

The result shows a shift in the time origin does not change the modulus of the spectrum. Only the phase is changed. If we are trying to calculate the modulus of a spectrum, we can locate the origin wherever we wish. In particular, for the case of a signal which has an axis of symmetry, it will be advantageous to take the time corresponding to the axis of symmetry as the origin, this will simplify the integration. This constancy of the modulus during a shift in t facilitates measurement of the power spectrum $|S(f)|^2$, of the signal s, since the signal $s(t)$ can be referred to any time origin.

Another consequence of this property is that in order to know the position of a given signal on the time axis, it is important to know the phase of its spectrum. To give only the phase is completely inadequate. The required definition will be seen in the following example.

Let us assume that a particularly simple seismic trace is composed of a symmetrical pulse centered at time t_0. If we wish to recover the trace from its spectrum, with what precision should the phase be known? The solution is very simple for we know that to within a multiple of π the phase is $\varphi = -2\pi f t_0$. From this we infer that the relative error in φ is equal to that incurred in t_0. If $t_0 = 2$ sec and $\Delta t_0 = 2$ ms, $\Delta\varphi/\varphi = 10^{-3}$.

Inasmuch as the position of the reflections on a given trace depends on their phase, it is important that the relative phases be conserved during filtering. The result demonstrated in the two previous assignments corresponds to the statement: when $s(t)$ is symmetrical and the time is measured from the axis of symmetry, the phase is zero or equal to a multiple of π sgn f (sgn $f = 1$ if $f > 0$ and -1 if $f < 0$). For example, if $s(t)$ has zero phase, $-s(t)$ has a phase equal to π sgn f or an odd multiple of this quantity. Furthermore, a symmetrical signal $s(t)$ becomes $s(t - t_0)$ if it is displaced so that its axis of symmetry is located at t_0, and on account of the translation theorem, the phase takes the form $p\pi$ sgn $f - \omega t_0$, p being a whole number. The phase has an ordinate at the origin which is a multiple of π sgn f and its slope immediately gives the displacement t_0 (Figure 4).

It is thus seen that if, by some procedure, the phase φ of a signal can be transformed to $\varphi - \omega t_0$, this signal will have been delayed by t_0 without changing its shape. This is done in electrical delay lines sometimes used for filtering.

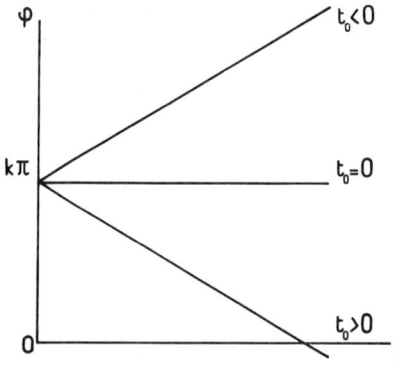

Figure 4.   Phase of symmetrical signal centered at t_0 as a function of frequency.

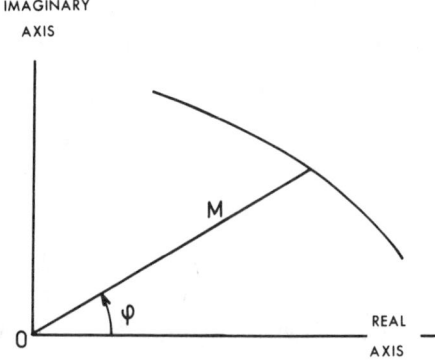

Figure 5. Locus of points in complex space generated by a vector of magnitude M and phase angle φ.

Such a delay line is made up of identical cells, each outputting a signal which, apart from the phase, is equal to that of the input. For example, the phase can be reduced by 0.2×10^{-3} ω radians, ω being expressed in radians per second, which corresponds to a delay of 0.2 msec. An appreciable delay is obtained by connecting a large number of such cells in series. It is, in fact, impossible to obtain circuits which do not alter the modulus of the signal spectrum and yet modify the phase strictly by ωt_0. Towards the high frequencies, in particular, dephasing may be less than ωt_0 and the modulus may be somewhat lowered. As a result, large delays cannot be imparted to signals whose spectrum rises too high in frequency.

In general, functions for which the spectrum is taken are not symmetrical and their phase no longer has so simple an aspect as described above. It is useful to know some general rules about the influence that complexity of the phase spectrum can have on the behavior of the corresponding function.

An initial comment may be made which stems directly from the inherent nature of the spectrum: For a given frequency, the spectrum is a complex number $Me^{j\varphi}$ that can be represented by the locus of its image in the complex plane.

A curve is thus obtained (Figure 5) which is traveled by the point (M, φ) in keeping with the frequency variations. This curve can be scaled in frequency. According to its geometry, it is seen that variations in the modulus and phase are not independent. In particular, when the line joining the origin to a point of the curve is tangential to this curve (Figure 6), φ is stationary and M on the contrary generally has a more rapid variation than at surrounding points. Hence the rule: An extremum of the phase generally has a corresponding point of inflection in the modulus. It can also be said that an extremum of the modulus generally has a corresponding point of inflection in the phase. These two rules are not absolute. In order for them to be strictly true, it is necessary that the velocity with which the representative point travels the curve, when f varies, be approximately uniform in the vicinity of the φ extremum in the first case and likewise in the vicinity of the M extremum in the second.

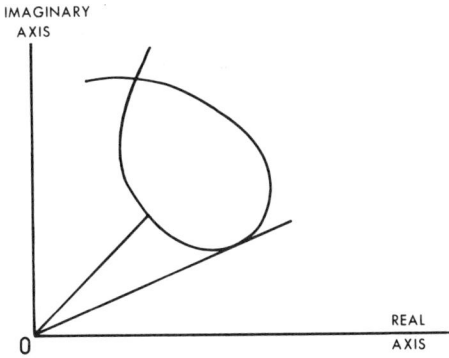

Figure 6. Locus of points in complex space generated by a vector of magnitude M and phase angle φ, showing the inverse relationship between the rates of change in M and φ.

Problem 10.

Assuming that the open circuit response of a velocity activated seismometer has a spectrum equal to:

$$\frac{p^2}{p^2 + 2\eta\,\omega_0 p + \omega_0^2}$$

where $p = j\omega$, $\omega_0 = 2\pi/T$, T *is the natural period and* η *the instrument damping, confirm that the inflection point rule holds good for* φ. *Find the phase corresponding to the points where the modulus is 3 db less than the maximum. Find the frequencies for which the dephasing equals* $\pi/4$ *and* $3\pi/4$. *Show that the greater* η *is, that is the more the seismometer is damped, the slower* φ *varies about the resonant frequency.*

Note that in regard to the last assignment, because here the spectrum is a rational fraction in powers of $j\omega$, there are some remarkable relations between modulus and phase that are not normally observed in the study of any spectra. When the modulus tends to a constant, the phase tends to zero. When the modulus varies as f^2, the phase is $\pm\pi$ sgn f. In general, just as is seen in the theory of mechanical and electrical filters, a modulus slope of $6n$ db/octave corresponding to a variation in the power n of f, is accompanied by a phase tending towards $\pm(n\,\pi/2)$ sgn f. It should not be forgotten, however, that the phase is only defined within a multiple of 2π, since it is the argument of a complex number. On the other hand, it should be borne in mind that a change of sign is equivalent to a dephasing by π sgn f, always within a multiple of 2π.

UNCERTAINTY RELATION. DELTA "FUNCTION"

It is known from the example of the seismometer response that the less it is damped, that is the more peaked the spectral modulus of the response, the longer the response. It is likewise known from the diffraction experiment of parallel, homogeneous light through a slit that the narrower the aperture, the broader is the diffraction pattern. This phenomenon is quite general and is known as the uncertainty principle.

Problem 11.

Generally speaking, if G(f) *is the spectrum of* g(t), *show that the spectrum of*

$$\boxed{g\,(t\,/\,t_0) = |t_0|\,G\,(f t_0)}$$

The result in Problem 11 indicates that if the scale of t is contracted or stretched, the scale of f behaves in an inverse sense. A short signal has a broad spectrum. Conversely, if the spectrum is narrow, the corresponding signal will be long; hence the name "uncertainty principle." If it is easy to locate a signal in t, it is difficult to do so in f and conversely. This will be confirmed in Problem 12.

Problem 12

A *"wave packet"* or *wave train s(t) is made up as follows:*

$$\begin{cases} s(t) = 0 & \begin{cases} t < 0, \\ t > T, \end{cases} \\ s(t) = \sin \omega_s t & 0 < t < T. \end{cases}$$

Calculate its spectrum. Find what the length of the wave train should be if we wish to determine its frequency with an accuracy of 1 percent, knowing that all frequencies are considered equally possible for the spectral modulus of s(t) extending between its maximum and 0.7 times this maximum (Figure 7).

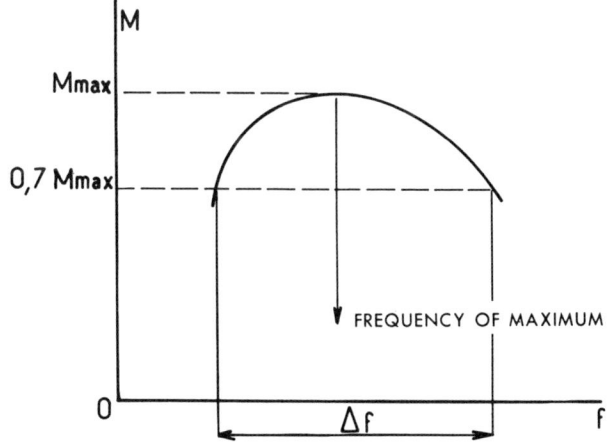

Figure 7. Spectrum of wave train and $s(t) = \sin \omega_0 t$ for $T \geqslant t > 0$ and $s(t)$ 0 elsewhere.

Problem 13

Calculate the spectrum for the square pulse g(t), g = 0 for $|t| > T/2$, g = 1 for $|t| < T/2$). If the difference in frequency for which the modulus is less than the maximum by 6 db is called the spectral width, find the relation which connects this width with T.

More generally, if the signal width Δt and that of the spectrum $\Delta\omega$ are suitably defined, it is demonstrated that $\Delta t \, \Delta\omega \geqslant 1$. But we must remain aware that Δt, in the case of a signal which oscillates many times, does not necessarily have an obvious physical meaning. Special reference will be made to the case of the signal s'' from the example given at the beginning of this chapter (Figure 3). Its spectrum extends from $-\infty$ to $+\infty$ with a constant modulus. We would then expect its width to be very small. As can be seen, such is not the case. Still, the uncertainty principle is not violated. In this respect it is sufficient to define Δt and $\Delta\omega$ correctly and to guard against intuitive notions of width and length of signals and spectra. The uncertainty principle is nevertheless valuable for studying signals of the same form and variable length. Therefore, as the reader will have shown, a scale change which divides time by t_0, that is which multiplies the length of $g(t)$ by t_0, will multiply frequencies by t_0 in $G(f)$, which corresponds to a contraction of the spectrum in dividing its width by t_0. The inequality has become an equality.

The most noted application of the relation $\Delta t \, \Delta\omega \geqslant 1$ is that made by Heisenberg (1927) from which it derives its name. He showed that the product of uncertainty Δy in the position of a particle and the uncertainty Δp in the momentum it possesses is greater than or equal to a universal constant. The basic demonstration of this theorem is readily done by employing on the one hand the laws of diffraction, and those of the Fourier transform on the other (Problem 14).

Problem 14

Knowing that a particle of mass m *with a rectilinear velocity* v *is associated with a plane wave of wavelength* $\lambda =$ h/mv *propagating in the direction of the vector* v, *if the particle is made to pass through a slit cut in a diaphragm perpendicular to* v *(Figure 8), show that the uncertainty* Δy *in the position of particle impact on a screen perpendicular to* v *is linked to the uncertainty* Δ(mv) *by the formula* $\Delta y \, \Delta$(mv) $\simeq$ 2h (h *is a universal constant*).

We see, after having completed this assignment, that there is nothing complicated or unexpected here. Philosophers surprised by the result, doubtless would have been much less so had they known about the properties of Fourier transformation. The difficulty lies more in the association of wave and particle.

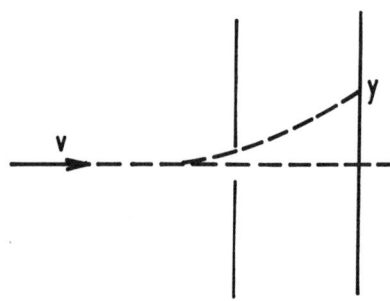

Figure 8.   Deflection of a particle by a slit.

Problem 15

If we accept for perception of music the value 2 for the product $\Delta f \Delta t$ of the spectral width and duration of a sound (f being in hertz and t in seconds), what is the inaccuracy in quarter tones of the A or "la" 3 note (440 hz) which last 100 msec (1/4 tone = 1/24 octave)?

In the same way, when a bundle of light is rapidly interrupted, its spectrum is slightly altered and becomes flatter as the light is allowed to pass in shorter trains.

Problem 16

A filter designed to study solar flares has a bandwidth of 4 Å about the hydrogen α line (λ=6563Å). What is the minimum duration of the train of optical vibrations passing through the filter? It will be assumed that the spectrum has the shape of a Gaussian curve whose Fourier transform is given below.

Let us take as another example the function $e^{-\theta^2}$ which is graphically portrayed by a bell-shaped Gaussian curve. If $\theta = t/t_0$ we can obtain a curve of variable width by making t_0 vary. The Gaussian curve represents one of the rare functions having the same shape as its spectrum. It will not be demonstrated here, for it is sufficient for us to know that the spectrum of $e^{-(t/t_0)^2}$ is $\sqrt{\pi}\,|t_0|\,e^{-(\omega t_0/2)^2}$. This function pair is interesting for more than one reason, and especially because it has only positive values. It is readily seen that the spectral modulus of the positive function $e^{-\theta^2}$ has a maximum for $f=0$. We shall make use of this to introduce the delta function concept which is fundamental to operational calculus and extremely convenient for the study of filters encountered in geophysics.

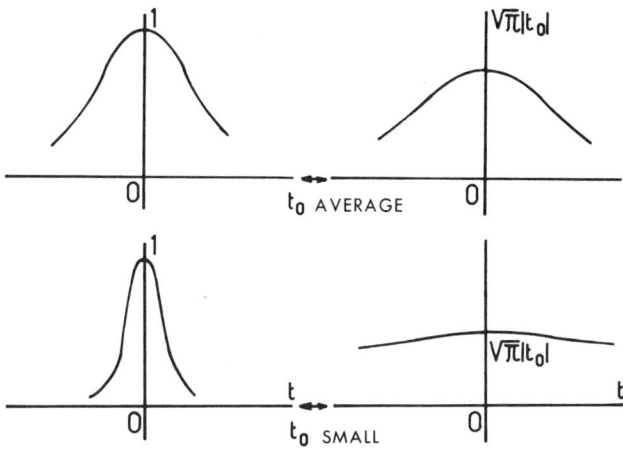

Figure 9.   Changes in shape and spectrum of the function $e^{-(t/t_0)^2}$ as t_0 changes.

Let us now vary t_0. If the curve $e^{-(t/t_0)^2}$ broadens, its spectrum narrows, and vice versa. We shall be able to confirm the principle of uncertainty on these curves by defining their width as the distance separating points of inflection. If t_0 becomes very small, the bell-curve function of time becomes very narrow and its spectrum very wide. As $t_0 \longrightarrow 0$, the curve $e^{-(t/t_0)^2}$ shrinks into the ordinate axis and the curve $\sqrt{\pi}\,|t_0|\,e^{-(\omega t_0/2)^2}$ spreads progressively, at the same time losing amplitude (Figure 9). Now, the area of the curve representative of $e^{-(t/t_0)^2}$ is equal to $\sqrt{\pi}\,|t_0|$.

If we now consider a new function equal to

$$\left[1/\left(\sqrt{\pi}\,|t_0|\right)\right]\,e^{-(t/t_0)^2},$$

it is seen that it also shrinks into the ordinate axis as t_0 becomes smaller, but that its area remains constant and equal to unity. One is tempted to extend this result to $t_0 = 0$, although the maximum of the function then becomes infinite. The spectrum which is equal to

$$e^{-(\omega t_0/2)^2},$$

spreads more and more and retains a constant maximum equal to 1 as t_0 diminishes. As a matter of fact, $G(0) = \int_{-\infty}^{\infty} g(t)\,dt$ which remains constant. In the limit, t_0 being zero, the time function tends to be represented graphically by a sort of needle[1] located at the abscissa zero; its height is infinite and its spectrum is a constant equal to unity (Figure 10). This is called the delta function $\delta(t)$ or Dirac function. Since it is exactly located, its spectrum is very broad.

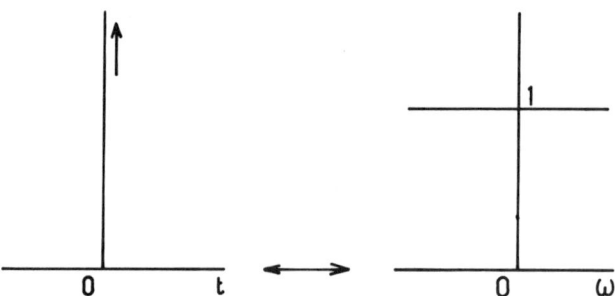

Figure 10.   Representation of the function $e^{-(t/t_0)^2}$ and its spectrum in the limit as $t_0 \to 0$.

It is clear from the foregoing that the delta function is not a function in the sense of classical analysis. It has in fact been defined as a limit.

This limit is everywhere zero (which is legitimate), except for $t = 0$ where

---

[1] It is interesting to note that the Germans call it the "needle function."

it is infinite (which is not legitimate). The δ function has been defined by passing to the limit where there is no limit. It is therefore not a function.

In order to obtain a correct mathematical definition of $\delta(t)$, the generalized function concept may be used (see, for example, Schwartz, 1961). This question will be studied in the chapter on density functions. Note that we would have been able to arrive at δ from other functions such as the square pulse or the function representing the discharge of a condenser. For this, it is sufficient to contract the functions while keeping their area constant.

The use that can be made of generalized functions in the theory of linear filters will be seen later. Until then, it will be noted that the delta function concept may already serve as a description of spectra which have been considered in the introduction. Let us return once more to the example in the physical introduction on the spectral concept. When (L) is occupied by sources vibrating in-phase and with the same amplitude, the amplitude of vibrations measured at P, a great distance away, is zero for all values of α different from zero (Figure 11). It may then be surmised that this amplitude can be expressed as a function of α in the form $u(0)\,\delta(\alpha)$, $u(0)$ being homogeneous with amplitude; and this is confirmed by the fact that the modulus of its spectrum is constant. When the amplitudes remain constant over (L), the phases being distributed sinusoidally, the amplitudes at great distance are zero for all values of α other than a certain α_0 fixed by the wavelength λ_0 measured over (L). The amplitude at great distance is then equal to $u(\alpha_0)\,\delta(\alpha - \alpha_0)$, $u(\alpha_0)$ being homogeneous with amplitude; this is in agreement with the theorem on the translation of the t origin. We have in fact demonstrated that if $S(f)$ is the spectrum of $s(t)$, $S(f)\,e^{-j\omega t_0}$ is the spectrum of $s(t - t_0)$. This corresponds in our case to the passage from $\delta(\alpha)$ to $\delta(\alpha - \alpha_0)$, the spectra of which are 1 and e^{-jk_0x}, x being the distance over (L) and k being equal to $1/\lambda_0$. By reasoning in this example of plane wave sources, the reader had already manipulated delta functions without knowing it. In order to be entirely rigorous, it would be necessary to say that the amplitude of the vibrations at P is proportional to $\delta(k - k_0)$ and not to $\delta(\alpha - \alpha_0)$. Clearly the physical meaning is the same if we know what the correct spectrum is.

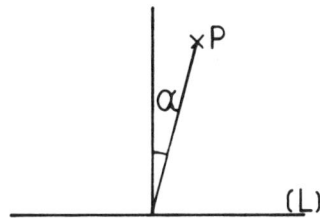

Figure 11. Point P situated at a great distance from a continuous line of sources vibrating in-phase.

DIFFERENTIATION AND INTEGRATION

In the theory of electrical circuits and mechanical systems, we have to deal with some derivatives and integrals of quantities such as current or velocity. If the velocity spectrum of a mass is known, how would we calculate, for example, the spectrum of the velocity imparted to a mass is known, how would we calculate, for example, the spectrum of the force exerted on it by a suspension spring? Here is an example of a problem whose solution is useful to the physicist. We shall see that such solutions go much further for they form the basis of a highly important technique in the study of all systems subject to linear differential equations.

$$g(t) = \int_{-\infty}^{\infty} G(f) e^{j\omega t} \, df,$$

The derivative of g will be

$$dg/dt = \int_{-\infty}^{\infty} G(f) j\omega \, e^{j\omega t} \, df.$$

The spectrum of dg/dt is therefore $j\omega\, G(\omega)$.

We infer from this that the spectrum of the integral of the function $g(t)$ is, within the spectrum of a constant C i.e., within $C\delta(f)$ equal to $G(f)/j\omega$.[2]

The first consequence is purely formal. We see that the operations of differentiation and integration correspond in the spectral domain to division or multiplication by $j\omega$. It follows that linear integral differential equations can be transformed to algebraic equations in $j\omega$ by simple Fourier transformation. We have achieved an operational calculus (Labin, 1949). Remember that a linear equation involves the function sought, its integrals and derivatives to the first degree only, products and other functions being excluded. The operational calculus in question applies only to equations with constant coefficients. Let us consider, with a view to future application to the theory of the moving coil seismometer, the example of a mass suspended at the end of a spring (Figure 12).

The classical equation $m\ddot{x} = -Kx$ transforms to $m(j\omega)^2 X = -KX$ where X is the Fourier transform of $x(t)$. We infer directly from this that the frequency can take only a single value:

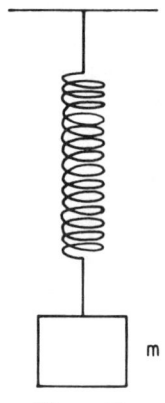

Figure 12.
Mass suspended
on a spring.

$$f = \frac{1}{2\pi} \sqrt{K/m}.$$

The motion of m is therefore sinusoidal. Its amplitude and phase depend on the initial conditions. Had a force been applied to the mass, it too would have been transformed and the solution would then have depended on its spectrum.

---

[2]These operations obviously assume the existence of $G(f)$ and of the integrals.

Problem 17
    What is the velocity of the mass m *suspended from a spring of stiffness* K *if a force* $F_0 \sin(\omega_0 t + \varphi_0)$ *is applied to it? Solve by Fourier transformation.*

Problem 18
    Find the voltage supplied in open circuit by the coil of an electromagnetic seismometer when the casing is constrained to the ground displacement u(t).

Problem 19
    Which among the following equations are amenable to treatment by operational calculus:

$$1/y^2 + \frac{d}{dx}(3/y) = 2, \qquad\qquad \frac{d}{dy}(fy) = e^{-y^2},$$

$$xy + 3\int_0^x y\, dx' + x = 0, \qquad\qquad \int_{-\infty}^t y\, dt' + \frac{d^2y}{dt^2} = 0.$$

$$\frac{d^3t}{df^3} + 60f + 0.1\, t = 7,$$

It is obvious that systems of linear integral-differential equations can be re-solved in this way, and this constitutes one of the most attractive qualities of the operational calculus. It is therefore possible to calculate the voltages at the output of a complicated electrical circuit knowing the form of the voltage applied to the input.

A second result of the theorems on differentiation and integration is the further possibility of calculating the Fourier transforms of known functions. Here are some examples:

Problem 20
    Knowing that S(f) *is the transform of* s(t), *what is the transform of* t s(t)? *What is the spectrum of* $(e^{-a^2t^2})/t$?

Problem 21
    Calculate the spectrum of the unit step u(t) = 1 *for* t>0, u(t) = 0 *for* t<0, *by means of the integration theorem.*

Problem 22
    Is the derivative of a real and symmetrical function symmetrical or asymmetrical? Show this by spectral argument.

Problem 23

    Confirm the spectrum previously calculated for the triangle:

$$
\begin{bmatrix}
s\,(t) = 0 & \begin{cases} t < -\,T, \\ t > T, \end{cases} \\
s\,(t) = t + T & -\,T < t < 0, \\
s\,(t) = T - t & 0 < t < T.
\end{bmatrix}
$$

    If this triangle is differentiated twice, we obtain $\delta\,(t + T) - 2\,\delta\,(t) + \delta\,(t - T)$. *What is its spectrum?*

Problem 24

    What is the relationship between the spectrum obtained by optical diffraction from a variable density trace and the spectrum calculated mathematically from the same trace?

    Finally, the third consequence of the theorems on integration and differentiation is the most important in physics for it concerns the possibility of performing certain mathematical operations by analog means. When we differentiate an electrical voltage or the rotation of a shaft which represents some physical magnitude, we are multiplying its spectrum by $j\omega$. But the physical magnitudes are always distorted by noise. If the noise is distributed in the spectrum through the high frequencies, these are going to be strongly amplified when multiplied by $j\omega$. There will then be more noise at the output of the differentiator than at the input, particularly for the high frequencies. It follows that, in an analog computer, differentiation is an operation prohibited from a practical point of view. It is of course completely out of the question to derive an acceleration from a displacement. It is always necessary, on the contrary, to calculate the displacement by double integration beginning with the acceleration.

Problem 25

    A system comprising an integrator and summation circuit establishes the following relation between the functions x(t) *and* y(t):

$$
\int y\, dt + x + y\,(1 - \alpha) = 0
$$

where α *is a positive or negative real number.*

    Write the relation which links X *and* Y, *the Fourier transforms of* x *and* y. *Sketch the spectrum and modulus of* Y/X *for* $\alpha = 0$, 1/2, 3/4, *and* 1. *State the conclusions concerning noise amplification at high frequency.*

Chapter 2

INTRODUCTION TO TWO-DIMENSIONAL SPECTRA

G. GRAU

Until now, only functions of a single variable have been considered. It is clear that the physical reasons which prompted us to introduce the concept of the single-dimension spectrum are also valid for two, three, four ... dimensions. Let us first look at the generalization in two dimensions. It is known that the spectrum of a linear source distribution $G(x)$ is the representation of a flux $g(\alpha)$ in a plane passing through the line (L) in question (Figure 2) according to the formula

$$g(\alpha) = \int_{-\infty}^{\infty} G(x)\, e^{2\pi jkx}\, dx \quad \text{with } k = (f \sin \alpha)/V.$$

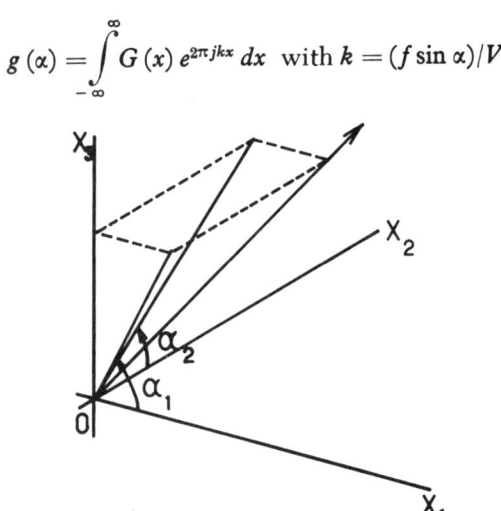

Figure 13. Direction to a distant source in three dimensions, rectangular coordinates.

If a plane distribution of sources is now considered, the flux that it generates is once again the sum of the contributions of the source elements. For a point located a great distance in the direction defined by the angles α_1 and α_2 (in a rectangular coordinate system $O\,x_1x_2x_3$) (Figure 13), the flux $g(\alpha_1, \alpha_2)$ is connected to the source distribution $G(x_1,x_2)$ in the $O\,x_1x_2$ plane by the relation:

$$g(\alpha_1, \alpha_2) = \int\int_{-\infty}^{\infty} G(x_1, x_2)\, e^{2\pi j(k_1x_1 + k_2x_2)}\, dx_1 dx_2 \quad .$$

As in the single-dimension problem, the k's are related to the α's by

$$k_1 = (f \sin \alpha_1)/V, \quad k_2 = (f \sin \alpha_2)/V.$$

Conversely, we will have

$$G\,(x_1, x_2) = \int\!\!\!\int\limits_{-\infty}^{\infty} g\,(k_1, k_2)\ e^{-2\pi j(k_1 x_1\,+\,k_2 x_2)}\ dk_1\ dk_2 \quad .$$

Here we have employed $g\,(\alpha_1, \alpha_2)$ and $g\,(k_1, k_2)$ indiscriminately in order to make the physical correspondence between G and g understood. Mathematically, the relation $g \longleftrightarrow G$ concerns $g\,(k_1, k_2)$ and $G\,(x_1, x_2)$.

Nothing is easier than to form spectra, or at least the square of their modulus, by an optical method which has previously been advocated in the study of single-dimension spectra.

It suffices that a monochromatic point source be situated at some distance and be viewed through a transparent film photo of a grating or a seismic section (with about twenty traces per millimeter). It will be adequate to adjust to infinity in order to see the spectrum. To photograph it, we focus it on a film (see Figure 43 of Chapter 5; see also the chapter on optical filtering).

We constantly must bear in mind the results of this experiment in what follows, for they permit us to understand clearly the properties of two-dimensional spectra more easily than would be possible by means of mathematical analysis of the formulas.

The double integrals may be calculated by two successive simple integrations. For example:

$$g\,(k_1, k_2) = \int\limits_{-\infty}^{\infty} e^{2\pi j k_1 x_1}\ dx_1 \int\limits_{-\infty}^{\infty} G\,(x_1, x_2)\ e^{2\pi j k_2 x_2}\ dx_2.$$

This means that the domain where the function G is different from zero may be cut into slices parallel to the x_2 axis. The spectrum of $G(x_1, x_2)$ is obtained by first taking the spectrum in k_2 corresponding to each of the bands and then taking the spectrum in k_1 of this first spectrum, or conversely. Since the contributions of all the elements of the domain should add up to $g(k_1, k_2)$, the way in which they are added matters little. One could as well sum on circular strips; it all depends on the nature of G and the shape of the contour which bounds the field of integration. We shall see this case later in the course of study of the functions with circular symmetry.

Let us now calculate the radiation of a square horn designed to provide an antenna with adequate directivity for linkage between two fixed stations following the hertzian relay technique (Figure 14).

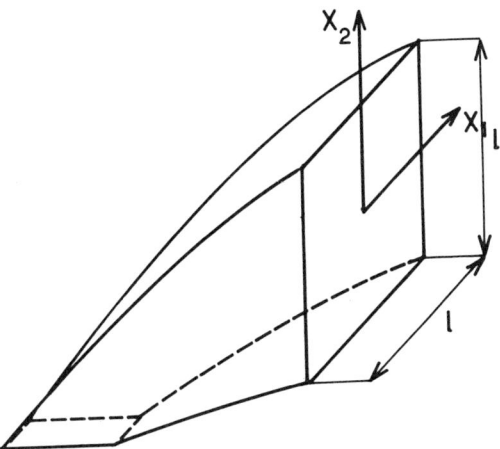

Figure 14.   Schematic representation of a horn.

It will be assumed that the wave reaching the opening is plane, its amplitude and phase being constant over the entire surface of the receiver:

$$\begin{cases} G\,(x_1, x_2) = 1 \\ G\,(x_1, x_2) = 0 \quad \text{elsewhere} \end{cases} \text{where} \begin{cases} |x_1| < 1/2, \\ |x_2| < 1/2, \end{cases}$$

We shall first calculate

$$\int_{-\infty}^{\infty} G\,(x_1, x_2)\, e^{2\pi j k_2 x_2}\, dx_2.$$

It is clear, whatever x_1 may be, that we are dealing with the spectrum of a rectangle. The integral becomes[3]

$$(\sin \pi l k_2)/(\pi k_2) = l\, \text{sinc}\,(l k_2)\,.$$

Next by integrating this quantity in x_1, after multiplication by $e^{2\pi j k_1 x_1}$, we finally have:

$$g\,(k_1, k_2) = l^2\, \text{sinc}\,(l k_2)\, \text{sinc}\,(l k_1)$$

$$= \frac{(\sin \pi l k_2)\,(\sin \pi l k_1)}{\pi^2\, k_2\, k_1}\,.$$

[3] Remembering that sinc $u = (\sin \pi\, u)/\pi\, u.$

The antenna is correspondingly more directive as l becomes larger. Since the double integral just calculated is separable, we obtain $g(k_1, k_2)$ in the form of a product of two functions, one of which depends only on k_1 and the other only on k_2.

This property is common to functions of two variables which can be expressed in the form of a product of two functions, each depending on one variable only. If $g(k_1, k_2) = g_1(k_1) g_2(k_2)$, and if $G_1 \leftrightarrow g_1$ and $G_2 \leftrightarrow g_2$ then $G_1(x_1) G_2(x_2) \leftrightarrow g_1(k_1) g_2(k_2)$.

CYLINDRICAL FUNCTIONS

By definition, these will be functions which vary only according to one of the variables. For example, $g(k_1, k_2)$ will be independent of k_2. In this case:

$$G(x_1, x_2) = \int\!\!\!\int_{-\infty}^{\infty} g(k_1, k_2) \, e^{-2\pi j(k_1 x_1 + k_2 x_2)} \, dk_1 \, dk_2$$

$$= \int_{-\infty}^{\infty} g(k_1, k_2) \, e^{-2\pi j k_1 x_1} \, dk_1 \int_{-\infty}^{\infty} e^{-2\pi j k_2 x_2} \, dk_2$$

$$= \delta(x_2) \int_{-\infty}^{\infty} g(k_1, k_2) \, e^{-2\pi j k_1 x_1} \, dk_1.$$

The spectrum of a cylindrical function (constant as a function of one of the variables) is zero for all nonzero values of the corresponding variable in the spectral domain. Where it is not zero, it is equal to the product of a delta function of this same variable with the one-dimensional spectrum of the generating function.

We may confirm this by optical diffraction. We see that by a certain arrangement, one-dimensional spectra are obtained with a point source without having to transform this to a linear source by means of a cylindrical lens.

FAN FUNCTIONS

By definition, this concerns functions which remain constant on all lines passing through the origin. Their form is therefore specific and should be expressible as a function of a single new variable equal to the ratio of the two original variables. Thus the fan function $g_i(k_1, k_2)$ of k_1 and k_2 may be written in the form $g_1(k_1/k_2)$. This property allows us to find easily the structure of the corresponding Fourier transform $G(x_1, x_2)$.

We shall use for this purpose the indeterminacy theorem as follows:

$$G(x_1, x_2) = \int\!\!\!\int_{-\infty}^{\infty} g_1(k_1/k_2)\, e^{-2\pi j(k_1 x_1 + k_2 x_2)}\, dk_1\, dk_2$$

$$= \int_{-\infty}^{\infty} e^{-2\pi jk_2 x_2}\, dk_2 \int_{-\infty}^{\infty} g_1(k_1/k_2)\, e^{-2\pi jk_1 x_1}\, dk_1.$$

The variable k_2 may be considered as a constant in the integral over k_1 and

$$\int_{-\infty}^{\infty} g_1(k_1/k_2)\, e^{-2\pi jk_1 x_1}\, dk_1 = |k_2|\, \Gamma_1(k_2 x_1),$$

where Γ_1 is the transform with one variable of g_1 that must, moreover, be carefully distinguished from the transform with two variables that we are seeking.

We therefore have:

$$G(x_1, x_2) = \int_{-\infty}^{\infty} |k_2|\, \Gamma_1(k_2 x_1)\, e^{-2\pi jk_2 x_2}\, dk_2.$$

The transform of $g_1(k_1/k_2)$ is the transform in k_2 of $|k_2|\, \Gamma_1(k_2 x_1)$, Γ_1 being the transform with one variable of g_1.

It is clear that we can also write $g(k_1, k_2)$ in the form $g_2(k_2/k_1)$, g_2 being in general different from g_1. We then find, in the same manner as above, that the transform of $g_2(k_2/k_1)$ is the transform in k_1 of $|k_1|\, \Gamma_2(k_1 x_2)$, Γ_2 being the transform with one variable of g_2.

We shall see in a subsequent chapter on spectra that G is written simply as a function of g itself and no longer of its transforms Γ_1 or Γ_2. In certain cases g_1 and g_2 may be identical. For example, if $g(k_1, k_2) = 0$ for $k_1/k_2 > 0$ and $g = 1$ for $k_1/k_2 < 0$, then $\Gamma_1 \equiv \Gamma_2$.

FUNCTIONS WITH CIRCULAR SYMMETRY

If $G(x_1, x_2)$ is unchanged by rotation about Ox_3, the calculation of $g(k_1, k_2)$ can be made very simply. In fact, by taking the optical example again, it is seen that any rotation about the Ox_3 axis should not alter the flux at the point of observation. As a result $g(k_1, k_2)$ should be unchanged by rotation. We should then be able to put $g(k_1, k_2)$ in the form of a function of $(k_1^2 + k_2^2)$ alone. If we denote this quantity by R^2 and the quantity $(x_1^2 + x_2^2)$ by r^2, we can write

$$x_1 = R \cos \alpha, \quad x_2 = R \sin \alpha, \quad k_1 = r \cos \beta, \quad k_2 = r \sin \beta.$$

We then have:

$$g\left(k_1, k_2\right) = \iint\limits_{-\infty}^{\infty} G\left(x_1, x_2\right) e^{2\pi j\left(k_1 x_1 + k_2 x_2\right)} dx_1 dx_2$$

$$= \int\limits_{0}^{\infty}\int\limits_{0}^{2\pi} G\left(R\right) e^{2\pi j r R \cos\left(\alpha-\beta\right)} R \, dR \, d\alpha,$$

$$g\left(r\right) = 2\pi \int\limits_{0}^{\infty} R \, G\left(R\right) J_0\left(2\pi r R\right) dR \quad ,$$

where $J_0\left(2\pi r R\right)$ is the Bessel function of zero order defined as

$$J_0\left(2\pi r R\right) = \frac{1}{2\pi} \int\limits_{0}^{2\pi} e^{2\pi j r R \cos u} \, du$$

(see e.g. Angot, 1957).

Conversely, we shall have:

$$G\left(R\right) = 2\pi \int\limits_{0}^{\infty} r \, g\left(r\right) J_0\left(2\pi r R\right) dr \quad .$$

These two integral relations between $g\left(k_1, k_2\right) = g\left(r\right)$ and $G\left(x_1, x_2\right) = G\left(R\right)$ are known under the name of Fourier-Bessel transformation and are widely used for the solution of problems of potential or of wave propagation in cylindrical symmetry. For such problems, in fact, this transformation plays the same role as the Fourier or Laplace transformation in single-dimension problems.

It will be observed, however, that the notations are not always the same. The factor 2π is sometimes absent. Here G is the spectrum of g in the sense adopted by us for the Fourier transformation. In any case, in our notation we shall have:

$$g\left(r\right) = 4\pi^2 \int\limits_{0}^{\infty} R \, J_0\left(2\pi r R\right) d R \int\limits_{0}^{\infty} r' \, J_0\left(2\pi r' R\right) g\left(r'\right) dr'$$

as with the single dimension:

$$g\left(t\right) = \int\limits_{-\infty}^{\infty} e^{j\omega t} \, d\omega \int\limits_{-\infty}^{\infty} e^{-j\omega t'} \, g\left(t'\right) dt'.$$

Problem 26

If the square horn for hertzian linkage is replaced by a circular horn of radius R_0, calculate the directivity function. It will be further assumed that the amplitude of the wave reaching the horn is constant. We shall make use of the identity

$$\frac{d}{du}[u^n J_n(u)] \equiv u^n J_{n-1}(u).$$

We can again, and always in the case of cylindrical symmetry, calculate the spectrum by means of a Fourier (and not Fourier-Bessel) integral with a single variable. As a matter of fact, because of the cylindrical symmetry of g, the integral

$$\int_{-\infty}^{\infty} G(x_1, x_2)\, e^{2\pi j k_2 x_2}\, dx_2$$

does not depend on k_2. We can make $k_2 = 0$, which gives

$$g(k_1, k_2) = \int_{-\infty}^{\infty} e^{2\pi j k_1 x_1}\, dx_1 \int_{-\infty}^{\infty} G(x_1, x_2)\, dx_2.$$

Owing to the symmetry of g and G, we can further write:

$$g(r) = \int_{-\infty}^{\infty} e^{2\pi j r x_1}\, dx_1 \int_{-\infty}^{\infty} G(x_1, x_2)\, dx_2.$$

g is the single dimension Fourier transform, with respect to one of the variables, of the G integral caluclated with respect to the other variable.

Readers interested in mathematics will be able to practice handling the problem of the circular section transducer in this manner.

Problem 27

Find the function $g(r)$ which corresponds to $G(R) = 1/R$.

Problem 28

Find the spectrum of $e^{-\pi r^2}$.

The form of the Fourier-Bessel integrals allows us to establish an interesting theorem on the value of $G(0)$. Let us assume that $g \longleftrightarrow G$. We then have:

$$G(R) = 2\pi \int_0^{\infty} r\, g(r)\, J_0(2\pi r R)\, dr.$$

We see immediately that

$$G\,(0) = 2\pi \int_0^\infty r\, g\,(r)\, dr \ .$$

Likewise

$$g\,(0) = 2\pi \int_0^\infty R\, G\,(R)\, dR \ .$$

It is easily verified that this is so in the case where $G(R) = 1$ for $R < R_0$ and 0 for $R > R_0$

Finally, to conclude these generalizations on spectra of functions with circular symmetry, the general formula corresponding to n-dimensional space is given:

$$G\,(R) = \frac{2\pi}{R^{(n-2)/2}} \int_0^\infty r^{n/2}\, g\,(r)\, J_{(n-2)/2}\,(2\pi r R)\, dr \ .$$

It is accompanied, of course, by the inverse formula permitting calculation of g from G.

For $n = 3$, we have

$$G\,(R) = \frac{2\pi}{R^{1/2}} \int_0^\infty r^{3/2}\, g\,(r)\, J_{1/2}\,(2\pi r R)\, dr$$

$$= \frac{2}{R} \int_0^\infty r\, g\,(r)\, \sin\,(2\pi r R)\, dr,$$

on account of the relation $J_{1/2}\,(u) = (2/\pi u)^{1/2} \sin u$.

If $\gamma(R)$ is the Fourier transform with one variable of $g(r)$, say $\int_{-\infty}^\infty g\,(r)\, e^{2\pi j r R}\, dr$, we have:

$$G\,(R) = (-\,1/2\pi R)\,(d\gamma/dR) \ \text{ since } \ \gamma\,(R) = 2 \int_0^\infty g\,(r)\, \cos 2\pi r R\, dr.$$

DELTA FUNCTION

As in single-dimension space, a delta function will be defined by a kind of limit of the function which is constant within a circle (or rectangle) and zero outside it, provided that its amplitude remains inversely proportional to the area maintained. This function will in fact be a density function.

Thus, in the case of the circle of radius R_0, G will be given an amplitude, not of 1, but of $1/(\pi R_0^2)$. The transform g is:

$$\frac{1}{\pi r R_0} J_1 (2\pi r R_0).$$

This tends to 1 as $R_0 \rightarrow 0$.

Here, as for the single-dimension case:

$$\boxed{\delta (R) \longleftrightarrow 1}.$$

We can again write:

$$\boxed{\delta (R) = \delta (x_1) \cdot \delta (x_2)}.$$

GENERAL PROPERTIES

Displacements by translation are handled in exactly the same way as with single-dimension problems. For example, for two dimensions:

if $g \longleftrightarrow G$, $$\boxed{g (x_1 - a, x_2 - b) \longleftrightarrow G \, e^{-2\pi j \, (k_1 a + k_2 b)}}.$$

We see here again that the modulus of the g spectrum is always insensitive to translations imposed on this function. This is readily ascertained by optical diffraction. Spectra obtained by this method are either observed with the naked eye or photographed. In both cases, all phase information is lost: only the modulus, or its square, is seen. But, it is easily observed that the diffraction image, that is to say the modulus, does not change when the object is displaced in its plane by translation.

As to scale changes of the function, they give rise to an inverse scale change in the spectral domain, similar to the single-dimension case. Two-dimension example:

if $g \longleftrightarrow G$, $$\boxed{g (x_1/a, x_2/b) \longleftrightarrow |ab| \, G (ak_1, bk_2)}.$$

This can be verified optically by preparing different scale reductions of the same object and viewing their diffraction figures (we then have $a = b$). We can make a vary independently of b by tilting the object about an axis parallel to that of x_2.

Rotations, on the other hand, are something new which do not exist in the case of single-dimension functions. If we consider the case of optical diffraction, we readily accept that if the object is rotated through a certain angle, the spectrum rotates by the same amount.

Let us find the law of transformation mathematically in the most general case of a linear change of coordinates. When a change of axis is made, the Fourier transform $G (k_1, k_2, ..., k_n)$ of $g (x_1, x_2, ..., x_n)$ assumes the new form $G' (k_1, k_2, ..., k_n)$ that is going to be inferred from G. In order to find the law of transformation, we shall generalize the method which enabled us to show that since the transform of $g(t)$ is $G(f)$, that of $g(at)$ is: $(1/|a|) G (f/a)$.

Let us now assume that we pass from the $x_1, x_2, ..., x_n$ coordinate system to the system $x'_1, x'_2, ... , x'_n$ retaining the same origin; the Fourier transform of $g(X)$ is changed to that of $g'(X')$. The notation X (or X') relates to the vector whose origin is the origin of the axes, and whose extremity is the point with coordinates $x_1, x_2, ..., x_n$ (or $x'_1, x'_2, ... , x'_n$). The notation $g(X)$ employed for simplicity denotes none other than $g (x_1, x_2, ..., x_n)$. The same will hold true for $G(K)$ which by definition will represent $G (k_1, k_2, ..., k_n)$. We therefore have:

$$G (K) = \int_{-\infty}^{\infty} ... \int_{-\infty}^{\infty} g (X) e^{-2\pi j(K \cdot X)} dv$$

In this integral, $K \cdot X$ and $K \cdot X'$ are the scalar product of vectors with n dimensions K and X; in matrix notation the K vector is a row vector and X is a column vector; dv and dv' are the differentials, small elemental volumes of two spaces with n dimensions, respectively equal to $dx_1 dx_2 ... dx_n$ and $dx'_1 dx'_2 ... dx'_n$. When X is changed to $X' = MX$, M being the square transformation matrix,

$g(X)$ is changed to $g(MX)$, and
$K \cdot X$ is changed to $K \cdot MX$,
and dv to $dv' = |M| dv$,

$|M|$ being the absolute value of the M determinant (see Valiron, 1948, p. 275 and 307). Arsac (1961) for his part, neglects to take the absolute value of the determinant in his general formula (p. 118).

The new transform is

$$G' (K) = \int_{-\infty}^{\infty} ... \int_{-\infty}^{\infty} g (MX) e^{-2\pi j(K \cdot X)} dv$$

which can be further written as

$$(1/|M|) \int_{-\infty}^{\infty} ... \int_{-\infty}^{\infty} g (M X) e^{-2\pi j(K M^{-1} \cdot M X)} |M| dv,$$

The integral has exactly the same form as

$$\int_{-\infty}^{\infty} ... \int_{-\infty}^{\infty} g (X) e^{-2\pi j(K \cdot X)} dv,$$

if the vector K is changed to KM^{-1}.

We can then say that:

> the transform of $g(MX)$ is $|M|^{-1} G(KM^{-1})$.

This formula obviously applies to cases of changes in the single-dimension variable. More specifically, to change t to at in the function $g(t)$ amounts to dividing the measurement unit of t by a. The matrix for the change of variable is then $M = a$, whose determinant is a. We than have $|M|^{-1} G(KM^{-1}) = |a|^{-1} G(k/a)$.

Assume now that it is a question of rotation in a two-dimension problem. We know that in this case $|M| = 1$. The formula then becomes $g(MX) \leftrightarrow G(KM^{-1})$, which, on account of the properties of rotation matrices, can be written $G(MK)$ if K is now considered as a column vector. We can say that the k's are transformed in the same way as the x's, and that G does not undergo change.

Problem 29
Establish this result directly for a rotation in two-dimensional space.

SYMMETRY

Let us again take the general case for functions of two variables, we have:

$$G(x_1, x_2) = \int\!\!\!\int_{-\infty}^{\infty} g(k_1, k_2) \, e^{-2\pi j(k_1 x_1 + k_2 x_2)} \, dk_1 \, dk_2.$$

Assume that g is entirely real and let us see what happens when we simultaneously change x_1 to $-x_1$ and x_2 to $-x_2$. It is seen that the real part of G is unchanged and that the imaginary part undergoes change of sign. On the whole, the modulus of G does not vary and the phase changes sign. In effect, all the spectra obtained by photography with optical diffraction apparatus are symmetrical with respect to the origin. Remember that these spectra are independent of the phase.

If $g(k_1, k_2)$ is real and symmetrical with respect to the origin, its spectrum G is real. In fact:

$$G(x_1, x_2) = \int\!\!\!\int_{-\infty}^{\infty} g(k_1, k_2) \cos 2\pi (k_1 x_1 + k_2 x_2) \, dk_1 \, dk_2$$

$$-j \int\!\!\!\int_{-\infty}^{\infty} g(k_1, k_2) \sin 2\pi (k_1 x_1 + k_2 x_2) \, dk_1 \, dk_2.$$

When we change k_1 to $-k_1$ and k_2 to $-k_2$ simultaneously, the quantity $g \sin 2\pi (k_1 x_1 + k_2 x_2)$ changes sign, which shows that the second integral is zero. The transform G is therefore real and reduced to the first integral. If $g(k_1, k_2)$ is real and antisymmetrical with respect to the origin, that is to say if

$g(k_1, k_2) = -g(-k_1, -k_2)$, it can be shown by similar reasoning that its transform $G(x_1, x_2)$ is imaginary.

Let us now search for the condition that $G(x_1, x_2)$ is even in x_1. We can write:

$$G(x_1, x_2) = \int_{-\infty}^{\infty} e^{-2\pi j k_2 x_2} \, dk_2 \int_{-\infty}^{\infty} g(k_1, k_2) \, e^{-2\pi j k_1 x_1} \, dk_1.$$

It is seen that x_1 appears only in the second integral. In order for G to be even in x_1, it is necessary and sufficient that this integral in k_1 be likewise. But, for this, it is necessary and sufficient to involve only the cosine of $2\pi k_1 x_1$, which can only happen if $g(k_1, k_2)$ is even in k_1. Hence the condition: for $G(x_1, x_2)$ to be even in x_1, it is necessary and sufficient that $g(k_1, k_2)$ be even in k_1.

Likewise it would be shown that for $G(x_1, x_2)$ to be odd in x_1, the necessary and sufficient condition is that $g(k_1, k_2)$ be odd in k_1.

In this last case, the value of the integral in k_1 will be found as:

$$-2j \int_{0}^{\infty} g(k_1, k_2) \sin(2\pi k_1 x_1) \, dk_1.$$

It follows that, if $g(k_1, k_2)$ is odd in k_1 and k_2 simultaneously, $G(x_1, x_2)$ is real, since the factor j vanishes by combining with the one that arises from the integration in k_2. Thus we again find a special case of the first of the theorems revealed in this study of symmetry.

We had, in fact, stated it in the reverse sense: it was g that was real and G which was symmetrical with respect to the origin. On the other hand, as we have the theorem: if $G(x_1, x_2)$ is the Fourier transform of $g(k_1, k_2)$, it is $g(-x_1, -x_2)$ which is the transform of $G(k_1, k_2)$, we can state all the theorems on symmetry in the opposite sense without further demonstration.

If g is odd in k_1 but even in k_2, G is imaginary. This is a special case of the theorem on antisymmetrical g.

Even if $g(k_1, k_2)$ is arbitrary, the spectrum G assumes noteworthy values along the axes. In fact, let us for example make $x_2 = 0$:

$$G(x_1, 0) = \iint_{-\infty}^{\infty} g(k_1, k_2) \, e^{-2\pi j k_1 x_1} \, dk_1 \, dk_2$$

$$= \int_{-\infty}^{\infty} e^{-2\pi j k_1 x_1} \, dk_1 \int_{-\infty}^{\infty} g(k_1, k_2) \, dk_2.$$

Along the x_1 axis, we shall therefore have the Fourier transform in k_1 of the integral in the direction k_2 of g.

We will be able to compare this result with that obtained above in the case of functions with circular symmetry. One should now review the case of the rectangular section transducer and that of cylindrical functions in light of this discussion.

The previous result is moreover useful in spectrography by optical diffraction for it allows us to obtain the average spectrum of traces in a section. It is sufficient to place the microfilm of the section in the apparatus and observe the light intensity on the frequency axis. We find here in fact the spectrum of the sum of traces without having to perform this summing.

If we now consider what ought to be seen on the other axis, it will be easy to convince ourselves that, mathematically speaking, nothing should be seen since each trace has a zero mean. However, due to the photographic representation of the microfilm, which does not accept negative values, it is more likely a constant that should be observed.

We shall find applications of these results in the chapter on signal and noise, the chapter on fan filters and in the chapter dealing with optical filtering.

Chapter 3

SOLUTIONS FROM THE CHAPTERS ON INTRODUCTION TO SPECTRA

G. GRAU

Problem 1 (p. 6)

If $G(f)$ is the spectrum of $g(t)$:

$$g(t) = \int_{-\infty}^{\infty} G(f) e^{j\omega t} \, df.$$

By placing this integral in the form:

$$\int_{-\infty}^{\infty} G(f) e^{-j\omega(-t)} \, df,$$

the spectrum of $G(f)$ is made to appear and we see that it is $g(-t)$ since:

$$g(-t) = \int_{-\infty}^{\infty} G(f) e^{-j\omega t} \, df.$$

Problem 2 (p. 6)

The function $g(t)$ has the spectrum:

$$G(f) = \int_{-\infty}^{\infty} g(t) e^{-j\omega t} \, dt = \int_{0}^{T} e^{-j\omega t} \, dt$$

$$= \int_{0}^{T} \cos \omega t \, dt - j \int_{0}^{T} \sin \omega t \, dt \;=\; \frac{1}{\omega} [\sin \omega T + j (\cos \omega T - 1)].$$

We can put it in yet another form, since:

$$G = \frac{1}{\omega} \left[2 \sin \frac{\omega T}{2} \cos \frac{\omega T}{2} - 2j \sin^2 \frac{\omega T}{2} \right]$$

$$= \frac{2}{\omega} \sin \frac{\omega T}{2} \left(\cos \frac{\omega T}{2} - j \sin \frac{\omega T}{2} \right)$$

$$= \frac{2}{\omega} \sin \frac{\omega T}{2} \, e^{-j\omega T/2}$$

$$= T \, \mathrm{sinc} \, (fT) \, e^{-j\omega T/2},$$

with the notation: $\mathrm{sinc} \, u = (\sin \pi u)/(\pi u)$

The modulus M is equal to $T |\text{sinc} (f T)|$. To calculate the proposed integral with a view to attempting synthesis of the square wave, it is necessary to decompose the frequency domain into elemental domains where the sinc function retains the same sign. Furthermore, we shall make use of the parity of M:

$$\int_{-\infty}^{\infty} M \, e^{j\omega t} \, df = 2 \int_{0}^{\infty} M \cos \omega t \, df$$

$$= 2 \left[\int_{0}^{1/T} + \int_{1/T}^{2/T} + \int_{2/T}^{3/T} + \ldots \right]$$

$$= 1/\pi \left\{ \left[\text{Si} \left(\pi \left(T + 2t \right) f \right) + \text{Si} \left(\pi \left(T - 2t \right) f \right) \right]_{0}^{1/T} - \left[\quad \right]_{1/T}^{2/T} + \ldots \right\}$$

since:

$$M \cos \omega t = \left| \frac{\sin \pi f T}{\pi f} \right| \cos \omega t$$

$$= \pm \frac{1}{2\pi f} \left[\sin \pi \left(T + 2t \right) f + \sin \pi \left(T - 2t \right) f \right],$$

with the $+$ sign if the sinc function is positive, $-$ sign in the opposite case. The notation Si represents the integral of the sinc function (see Boll, 1957, p. 366-369):

$$\text{Si} \left(u \right) = \int_{0}^{u} \frac{\sin v}{v} \, dv.$$

From the modulus alone, we then finally recover:

$$\frac{2}{\pi} \left[\text{Si} \left(\pi \frac{T + 2t}{T} \right) + \text{Si} \left(3\pi \frac{T + 2t}{T} \right) + \text{Si} \left(5\pi \frac{T + 2t}{T} \right) + \ldots \right.$$

$$+ \text{Si} \left(\pi \frac{T - 2t}{T} \right) + \text{Si} \left(3\pi \frac{T - 2t}{T} \right) + \text{Si} \left(5\pi \frac{T - 2t}{T} \right) + \ldots$$

$$- \text{Si} \left(2\pi \frac{T + 2t}{T} \right) - \text{Si} \left(4\pi \frac{T + 2t}{T} \right) - \text{Si} \left(6\pi \frac{T + 2t}{T} \right) - \ldots$$

$$\left. - \text{Si} \left(2\pi \frac{T - 2t}{T} \right) - \text{Si} \left(4\pi \frac{T - 2t}{T} \right) - \text{Si} \left(6\pi \frac{T - 2t}{T} \right) - \ldots \right].$$

It is easy to see that this function cannot be equal to the original g function, if only because of its parity in t.

Problem 3 (p. 7)

We see that

$$s(t) = T - |t| \text{ between } -T \text{ and } +T.$$

$$S(f) = \int_{-\infty}^{\infty} s(t) \cos \omega t \, dt - j \int_{-\infty}^{\infty} s(t) \sin \omega t \, dt.$$

Since $s(t)$ is even, $s(t) \sin \omega t$ is odd, implying

$$\int_{-\infty}^{\infty} s \sin \omega t \, dt = 0$$

and

$$S(f) = \int_{-\infty}^{\infty} (T - |t|) \cos \omega t \, dt = 2 \int_{0}^{T} (T - t) \cos \omega t \, dt.$$

Now

$$- T \int_{0}^{T} \cos \omega t \, dt = T \, \frac{\sin \omega T}{\omega}$$

and

$$\int_{0}^{T} t \cos \omega t \, dt = \frac{1}{\omega^2} \int_{0}^{\omega T} u \cos u \, du \quad \text{(with } u = \omega t\text{)}$$

$$= T \, \frac{\sin \omega T}{\omega} + \frac{\cos \omega T - 1}{\omega^2}.$$

Hence

$$S(f) = 2T \, \frac{\sin \omega T}{\omega} - 2 \left[T \frac{\sin \omega T}{\omega} + \frac{\cos \omega T - 1}{\omega^2} \right]$$

$$= 2 \, \frac{1 - \cos \omega T}{\omega^2} = \frac{4}{\omega^2} \sin^2 \frac{\omega T}{2}$$

$$T^2 \, [\text{sinc} \, (fT)]^2.$$

Problem 4 (p. 7)

Let $s(t)$ be a real function of t.

Its transform $S(f)$ is $\int_{-\infty}^{\infty} s(t)\, e^{-j\omega t}\, dt$.

The real part of $S(f)$ is $R(f) = \int_{-\infty}^{\infty} s(t)\cos \omega t\, dt$.

$s(t)$ is independent of ω and $\cos \omega t$ is even in ω; therefore $R(f)$ is an even function of ω.

Similarly, the imaginary part $I(f) = -\int_{-\infty}^{\infty} s(t)\sin \omega t\, dt$ is an odd function of ω.

Problem 5 (p. 8)

If $s(t)$ is a real function of t,

its transform $S(f) = R(f) + j\,I(f)$

where $R(f) = $ real part of $S(f)$,

$\quad\quad I(f) = $ imaginary part of $S(f)$,

and can be put in the form $Me^{j\varphi}$

where $M = $ modulus of $S(f)$

$\quad\quad \varphi = $ phase of $S(f)$.

It is known that $R(f)$ is an even function of ω and that $I(f)$ is an odd function of ω.

$M = \sqrt{R^2 + I^2}$ is an even function of ω, since R^2 and I^2 are both even in ω.

$\varphi = \arctan I/R$ is an odd function of ω, since I/R is odd in ω.

Problem 6 (p. 8)

Let $g(t)$ be the real function in question:

$$g(t) = \int_{-\infty}^{\infty} (R + j\,I)\, e^{j\omega t}\, df.$$

$$= \int_{-\infty}^{\infty} (R\cos \omega t - I\sin \omega t)\, df + j\int_{-\infty}^{\infty} (R\sin \omega t + I\cos \omega t)\, df.$$

The second integral must be zero since $g(t)$ is real.

Furthermore, we can write that if $t > 0$, inasmuch as $g(t)$ is nonzero, $g(-t)$ is identically zero; that is:

$$0 = \int_{-\infty}^{\infty} (R\cos \omega t + I\sin \omega t)\, df.$$

By adding the two last equations, we infer that:

$$g(t) = 2\int_{-\infty}^{\infty} R\cos \omega t\, df,$$

and by subtracting them:

$$g(t) = -2 \int_{-\infty}^{\infty} I \sin \omega t \, df.$$

Because R is even and I is odd, we can again write:

$$g(t) = 4 \int_{0}^{\infty} R \cos \omega t \, df = -4 \int_{0}^{\infty} I \sin \omega t \, df.$$

These results are very important since the functions equal to zero for $t < 0$ are frequently met in physics. These will be, for example, any system response to an excitation starting at the time $t = 0$. In such a case, the real part of G, or the imaginary part, is sufficient to describe g. Furthermore, R and I are connected by the relation:

$$\int_{0}^{\infty} R \cos \omega t \, df = -\int_{0}^{\infty} I \sin \omega t \, df.$$

We shall see other forms of this equation further on.

Problem 7 (p. 8)

$$G(f) = \int_{-\infty}^{\infty} g(t) e^{-j\omega t} \, dt, \text{ which will be written } g(t) \longleftrightarrow G(f).$$

We look for the transform of $g(-t): g(-t) \longleftrightarrow \int_{-\infty}^{\infty} g(-t) e^{-j\omega t} \, dt.$

Setting $-t = u$:

$$g(u) \longleftrightarrow -\int_{\infty}^{-\infty} g(u) e^{j\omega u} \, du = \int_{-\infty}^{\infty} g(u) e^{-j(-\omega)u} \, du;$$

then: $g(-t) \longleftrightarrow G(-f).$

Let us assume that $G(-f) = G^*(f)$:

if $G(f) = R(f) + j I(f)$, $\begin{cases} G(-f) = R(-f) + j I(-f) \\ G^*(f) = R(f) - j I(f). \end{cases}$

Then this implies:

$$\begin{cases} R(-f) = R(f), \\ I(-f) = -I(f). \end{cases}$$

This amounts to saying that R is even and I is odd in ω.

Now:

$$g(t) = \int_{-\infty}^{\infty} (R + j I) e^{j\omega t} \, df.$$

By taking account of the even and odd relations:

$$g(t) = \int_{-\infty}^{\infty} [R(-f) - jI(-f)] e^{j\omega t} \, df.$$

If we set $-f = u$,

$$e^{j\omega t} = e^{-2\pi j u t}$$

and:

$$g(t) = -\int_{\infty}^{-\infty} [R(u) - jI(u)] e^{-2\pi j u t} \, du = \int_{-\infty}^{\infty} [R(u) - jI(u)] e^{-2\pi j u t} \, du.$$

This integral is equal to the expression of $g(t)$ where j would have changed to $-j$.

Therefore: $g(t) = g^*(t)$ and $g(t)$ is real.

Problem 8 (p. 9)

The equality $g(t_0 + t_1) = g(t_0 - t_1)$ is written as a function of the spectrum:

$$\int_{-\infty}^{\infty} G(f) e^{j\omega(t_0 + t_1)} \, df = \int_{-\infty}^{\infty} G(f) e^{j\omega(t_0 - t_1)} \, df.$$

It is obviously inferred from this:

$$0 = \int_{-\infty}^{\infty} G(f) [e^{j\omega(t_0 + t_1)} - e^{j\omega(t_0 - t_1)}] \, df.$$

If M is the modulus of G and φ is the phase, this equation can be put in the form:

$$0 = \int_{-\infty}^{\infty} M \, e^{j(\varphi + \omega t_0)} [e^{j\omega t_1} - e^{-j\omega t_1}] \, df$$

$$= 2j \int_{-\infty}^{\infty} M \, e^{j(\varphi + \omega t_0)} \sin \omega t_1 \, df.$$

For this to be true, the real part of this last integral:

$$-2 \int_{-\infty}^{\infty} M \sin(\varphi + \omega t_0) \sin \omega t_1 \, df$$

and its imaginary part:

$$2 \int_{-\infty}^{\infty} M \cos(\varphi + \omega t_0) \sin \omega t_1 \, df$$

must be zero for all t_1. It is necessary that $\varphi = -\omega t_0$, to within an integral multiple of π in order for the real part to be zero; the imaginary part must then be zero, owing to the parity of $M(f)$.

Because of the oddness of the spectral phase of a real function, the spectral phase of the function $g(t)$ considered will have the form:

$$\varphi = -\omega t_0 + p\pi \; \text{sgn} \, f,$$

with:

$$\text{sgn} \, f = \quad 1 \quad \text{if} \quad f > 0$$
$$= -1 \quad \text{if} \quad f < 0.$$

Problem 9 (P. 9)

Let $g(t)$ be a function whose spectrum is $G(f)$. Shifting the time origin by t_0 in the direction of positive time amounts to retaining the same origin and introducing a shift t_0 for the whole of the $g(t)$ function towards negative time. $g(t)$ is therefore transformed into $g(t + t_0)$.

The new spectrum is consequently:

$$\int_{-\infty}^{\infty} g\,(t + t_0)\, e^{-j\omega t}\, dt$$

$$= \int_{-\infty}^{\infty} g\,(t + t_0)\, e^{-j\omega(t + t_0)}\, e^{j\omega t_0}\, d\,(t + t_0)$$

$$= e^{j\omega t_0}\, G\,(f).$$

In order to apply this theorem to explain the previous problem, we take a function which is real and symmetrical relative to $t = 0$ as the initial function. We know that its phase is equal to $p\pi \; \text{sgn} f$ (p being an integer).

We pass from this function to the same function made symmetrical with respect to $t = t_0$ by displacement of the time origin of t_0 towards negative t. It is now necessary to apply the theorem just demonstrated by changing the term t_0 to $-t_0$. The phase then becomes $-\omega t_0 + p\pi \; \text{sgn} f$.

Problem 10 (p. 12)

We make the following changes of variable in the formula:

$$p = 2\pi j f \quad \text{and} \quad \omega_0 = 2\pi/T.$$

The spectrum being $-T^2 f^2/(-T^2 f^2 + 2j\eta Tf + 1)$, the modulus is:

$$M = T^2 f^2/[(1 - T^2 f^2)^2 + 4\eta^2 T^2 f^2]^{1/2},$$

and the phase:

$$\varphi = \text{Arctan} \, [2\eta Tf/(T^2 f^2 - 1)].$$

For the frequency $1/T$, the modulus has the value $1/2\eta$ (Figure 15a). The slope of the modulus curve is $T/(2\eta)$. It follows that, for each of the curves in Figure

15a, the tangent at the abscissa point 1 passes through the origin.

For the frequency $1/T$, the phase has the value $\pi/2$. Insofar as the derivative $d\varphi/df$ is equal to $-2\eta T (T^2f^2 + 1) / [(1 - T^2f^2)^2 + 4\eta^2 T^2f^2]$ the slope of the phase characteristic $\varphi(f)$ is equal to $-T/\eta$ at this abscissa (Figure 15b). At zero frequency, the spectrum is equivalent to $-T^2f^2$, the modulus tends to zero with a parabolic behavior and the phase tends to π with the slope $-2\eta T$.

At infinite frequency, the spectrum is equal to 1. There is therefore an asymptotic branch on the modulus curve with asymptote $M = 1$ and one on the phase curve with asymptote $\varphi = 0$.

The derivative of the modulus being:

$$2 T^2f [1 + T^2f^2 (2\eta^2 - 1)] / [(1 - T^2f^2)^2 + 4\eta^2 T^2f^2]^{3/2},$$

the modulus has a maximum at the frequency $1/(T \sqrt{1 - 2\eta^2})$ if $\eta < 1/\sqrt{2}$. The value of this maximum is:

$$1/2\eta \sqrt{1 - \eta^2}.$$

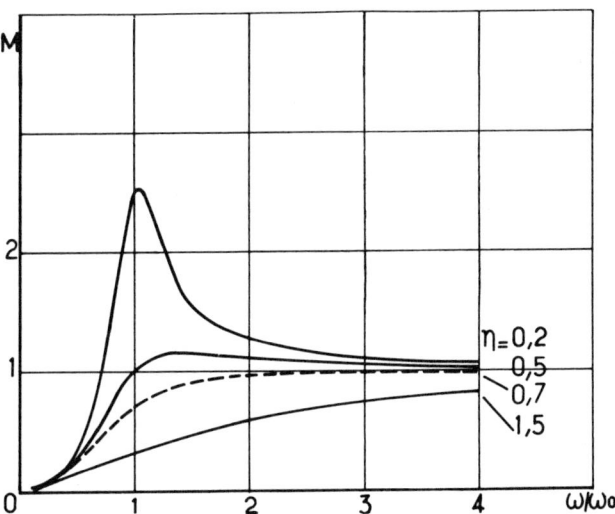

Figure 15a. Open circuit response of a velocity activated seismometer — modulus of spectrum.

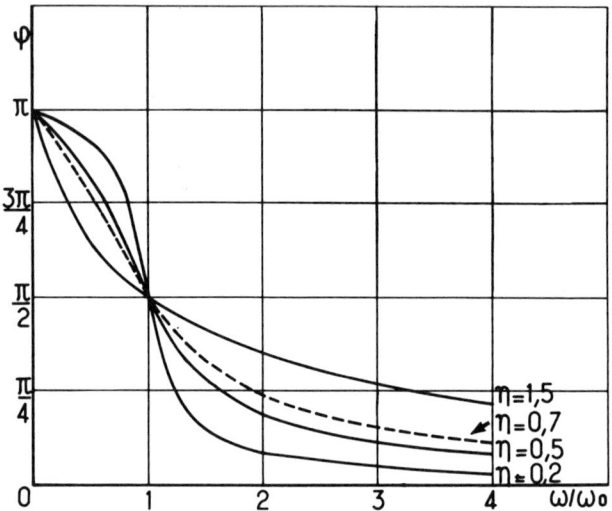

Figure 15b. Open circuit response of an activated seismometer — phase.

For values of η greater than $1/\sqrt{2}$, the modulus has no maximum; hence the name optimum value of η is given to $1/\sqrt{2} = 0.707$ (dashed curves).

We see that the inflection point rule is not honored. The maxima of the modulus lie to the right of the abscissa $Tf = 1$ and the inflection points of the phase are to the left. The situation would be much better from this point of view for the spectrum $1/(-T^2f^2 + 2j\eta Tf + 1)$. Likewise, the inflection points of the modulus do not correspond to any extremum of the phase, which does not moreover have any at finite frequencies.

If the phase is equal to $\pi/4$ or to $3\pi/4$, we have

$$2\eta Tf / (T^2f^2 - 1) = \pm 1,$$

hence $Tf = \pm \eta + \sqrt{\eta^2 + 1}$. The points are not located at abscissae symmetrical with respect to $Tf = 1$, but are closer to being so as η becomes smaller; for then $Tf \simeq 1 \pm \eta$. For example for $\eta = 0.2$, we have $Tf = 1.22$ and 0.78. These frequency values are useful, at least for the case where η is small, since the modulus M is close to $Tf/(2\sqrt{2}\eta)$ here; that is, if $\eta \ll 1$, an attenuation around 3 db relative to the maximum. This is in fact close to $Tf/(2\eta)$. It is easy to see on the curve corresponding to the value $\eta = 0.5$ that this rule is no longer valid at all when η is not very small. In this respect, 0.1 is approximately the limiting value. Likewise, the rule, which dictates that the points where M is half-maximum are located at frequencies $Tf \pm \eta \sqrt{3}$, is valid only for very small η.

Problem 11 (p. 12)

Let $G(f) = \int\limits_{-\infty}^{\infty} g(t) e^{-j\omega t} dt$.

Calculate the transform of $g(t/t_0)$: $\int\limits_{-\infty}^{\infty} g(t/t_0) e^{-j\omega t} dt$.

Make change of variable $t/t_0 = u$.

If $t_0 > 0$, $g(u) \longleftrightarrow t_0 \int\limits_{-\infty}^{\infty} g(u) e^{-j\omega u t_0} du = t_0 G(ft_0)$.

If $t_0 < 0$, the limits are inverted by the change of variable:

$$g(u) \longleftrightarrow t_0 \int\limits_{\infty}^{-\infty} g(u) e^{-j\omega u t_0} du = -t_0 G(ft_0).$$

For any t_0, therefore, $g(t/t_0) \longleftrightarrow |t_0| G(ft_0)$.

Problem 12 (p. 13)

We shall assume for simplicity that the signal $s(t)$ comprises a whole number n of periods. We will then be able to change the origin so that $s(t)$ becomes odd. The spectrum $S'(f)$ of the signal $s'(t)$ obtained this way is equal to the product of the spectrum $S(f)$ of $s(t)$ with:

$$e^{j\omega T/2}.$$

If we set $t = t' + T/2$, $s(t)$ becomes odd. Since S' and S have the same modulus, and as we are eventually concerned only with the modulus, we shall calculate $|S'|$.

Because s' is odd, S' is entirely imaginary:

$$S' = \pm j \int\limits_{-T/2}^{T/2} (\sin \omega_0 t)(\sin \omega t)\, dt,$$

the positive sign applying in the case where n is odd, the negative sign when n is even.

Hence

$$|S'| = \left| 2 \int\limits_{0}^{T/2} (\sin \omega_0 t)(\sin \omega t)\, dt \right|$$

$$= \left| \int\limits_{0}^{T/2} \cos(\omega + \omega_0)t\, dt - \int\limits_{0}^{T/2} \cos(\omega - \omega_0)t\, dt \right|;$$

$$|S'| = \left| \frac{1}{\omega + \omega_0} \left[\sin(\omega + \omega_0)t \right]_{0}^{T/2} - \frac{1}{\omega - \omega_0} \left[\sin(\omega - \omega_0)t \right]_{0}^{T/2} \right|.$$

Finally the modulus M of S equals

$$\frac{1}{\omega + \omega_0} \sin{(\omega + \omega_0)} \frac{T}{2} - \frac{1}{\omega - \omega_0} \sin{(\omega - \omega_0)} \frac{T}{2} .$$

When T is large enough, M has maxima for values of ω essentially equal to $\pm \omega_0$. Figure 16 corresponds to the case $n = 3$.

Now $T = n T_0 = \frac{2\pi}{\omega_0} n$; when n is large the shape of the representative curve of M in the vicinity of $\pm \omega_0$ is that of $\frac{(\sin \omega T/2)}{\omega} = \frac{T}{2} \frac{(\sin \omega T/2)}{\omega T/2}$ (dotted and dashed in the diagram).

M then attains $0.7\ M_{max}$ for $\frac{\omega T}{2} = 1.4$; hence $\omega = \frac{2.8}{T} = \frac{2.8}{2n\pi} \omega_0$; (since $\frac{\sin x}{x} = 0.7$ for $x = 1.4$; see Boll, 1957, p. 327).

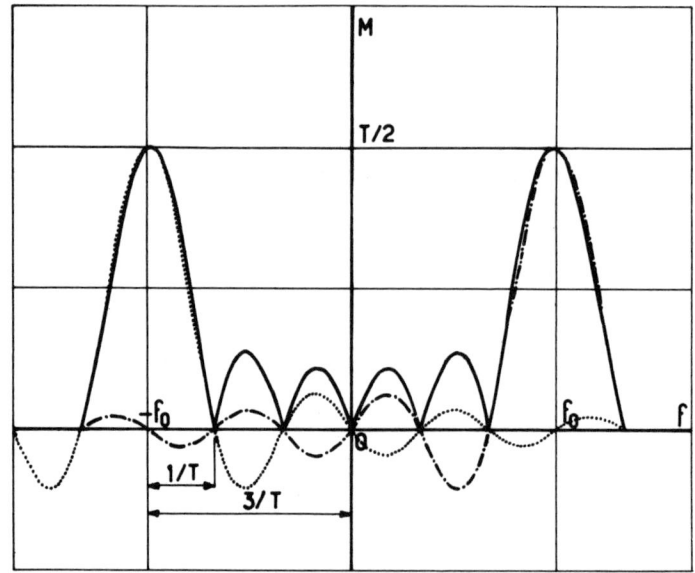

Figure 16. Amplitude spectrum of the wave train $s(t)$ equals $\sin \omega_0 t$, $0 < t < T$ where T corresponds to three whole wavelengths. Dotted and dashed lines represent same spectrum around $-f_0$ and f_0 for n large.

The difference $\Delta \omega$ is equal to $\left(\frac{2.8}{2n\pi} \omega_0 \right) \times 2$. It is desirable that the error in f_0 be less than 1 percent, whence we have $\frac{\Delta f}{f_0} = \frac{\Delta \omega}{\omega_0} = \frac{1}{100} = 2 \frac{2.8}{6.28n}$. Therefore

$n = \frac{560}{6.28} \simeq 90$. The signal should be about 90 periods in length.

The frequency f_0 is adequately defined only if the wave train is long enough. If we desire an accuracy of $1/p$, we must observe 0.9 p periods, which requires a measurement time equal to 0.9 p/f_0. In order to define a frequency of 25 hz within 2 percent, 45 periods are necessary, hence an observation interval of 1.8 sec. If a wave train of 15 hz lasts only 0.5 sec, the accuracy with which its frequency can be measured falls to 12 percent! This illustrates, among other things, the difficulty that would be encountered in filtering marine seismic data if we wished to estimate resonant frequencies in this manner. It is seldom, indeed, that the trains of oscillation last as long without interference. We would obviously obtain different values of n if other criteria for detecting the peak M were employed and especially if f_0 were estimated by the average of frequencies for which $M = 0.7\, M_{max}$.

Problem 13 (p. 13)

Because of the parity of $g(t)$, its spectrum can be written in the form:

$$G(f) = 2 \int_0^{T/2} \cos \omega t \; dt.$$

Hence

$$G(f) = \frac{2}{\omega} \sin(\omega T/2) = T \operatorname{sinc}(fT).$$

This result will be compared with that of a previous problem (Problem 2).

The sinc u function has a maximum equal to 1 for $u = 0$. It takes the value 1/2 for $\pi u = \pm 1.895$ (Boll, 1957, p. 329).

The width Δf is then $2(1.9/\pi T)$ and we have $T \cdot \Delta f = 1.21$.

It will be noted that sinc functions had already been obtained in solving the previous assignment. The reason for it will be given in the chapter on convolution. The relation which applies to the band defined by the 3 db points is $T \cdot \Delta f = 0.89$. In the case of 25 hz, we have $T = 1.8$ sec and $\Delta f = 0.5$ hz.

Problem 14 (p. 14)

Let d be the width of the slit cut in the diaphragm, $\alpha/2$ the half-angle aperture of the diffracted beam (central pattern) and λ the wavelength h/mv associated with the particles.

According to the law of diffraction, the central diffraction pattern caused by the slit with width d is such that:

$$\sin(\alpha/2) = \lambda/d.$$

Since α is small, we have approximately $\sin \alpha = 2\lambda/d$.

The momentum in the $0x$ direction is $p = h/\lambda$. On account of the diffraction, movement along $0y$ is uncertain by an amount:

$$(\Delta p)_y = \frac{h}{\lambda} \sin \alpha,$$

instead of remaining strictly zero.
    Therefore

$$(\Delta p)_y \simeq \frac{h}{\lambda}\frac{2\lambda}{d} = \frac{2h}{d}.$$

But d is equal to the uncertainty Δy in the initial position of the incident beam on the slit. Hence the relation:

$$\Delta y \cdot (\Delta p)_y \simeq 2h.$$

Problem 15 (p. 15)

The frequency difference Δf equals $2/\Delta t = 20$ hz. Therefore $\Delta f/f = 20/440 = 1/22$. The extreme frequencies that will be estimated will be:

$$440(43/44) \text{ and } 440(45/44)$$

The difference is therefore less than 1/4 tone ($\sqrt[24]{2} = 1.03$).

Problem 16 (p. 15)

We assume that the vibration train has the spectrum of a Gaussian curve centered at $\lambda = 6563$Å, that is 6563×10^{-8} cm. The spectrum will have approximately the same shape in the frequency domain since the 4 Å width is small in relation to the mean wavelength.

The spectrum being a bell-shaped curve is of the form $e^{-K(f-f_0)^2}$. According to the displacement theorem, the variation of light intensity with time has the form:

$$e^{j\omega_0 t}\, e^{-K't^2}.$$

It is obvious that a complex exponential should, as usual, be taken in the following sense: the variation is sinusoid (frequency $\omega_0/2\pi$) with a fixed phase which remains to be determined. As we are interested only in the duration of the wave train, we shall work exclusively with the envelope in the form of a Gaussian curve, which is the transform of the Gaussian curve by which we describe the passband of the filter (provided that the latter is referred to an origin located on its axis of symmetry).

Let us look for the uncertainty relation corresponding to the Gaussian curves by taking the distance separating the inflection points as the width. These correspond to an ordinate equal to 60 percent that of the peak.

Let us accept that

$$e^{-(t/t_0)^2} \longleftrightarrow \sqrt{\pi}\,|t_0|\,e^{-(\omega t_0/2)^2}.$$

The inflection points have abscissae $t = \pm t_0/\sqrt{2}$ and $f = \pm 1/(\sqrt{2}\,\pi t_0)$. The product of the uncertainties is therefore:

$$\Delta t \cdot \Delta f = 2/\pi = 0.637.$$

Now $f = c/\lambda$; hence $\Delta f = c\Delta\lambda/\lambda^2$ and $\Delta t = 0.637\,\lambda^2/c\,\Delta\lambda$.

That is: $\Delta t = 2.28 \times 10^{-12}$ sec $(c = 3 \times 10^{10}$ cm/sec$)$.

This time may be expressed as a number of periods. The number is $n = \Delta t/(\lambda/c)$ $= 0.637\,\lambda/\Delta\lambda \simeq 1040$. This result may be compared with that of the assignment where we attempted to measure a frequency to within 1 percent (see Problem 10).

During the time Δt, the light will have traversed $\Delta l = c\Delta t = n\lambda = 6.8\times10^{-2}$cm. Whatever the coherence of the light transmitted to the filter may be, the light leaving it (if there is any) is coherent over a distance along the beam equal to about 7/10 of a millimeter.

It will be noted that the formula which gives Δt may be put in the form:

$$\Delta t = C\,Q\,\lambda/c$$

where: C is a constant depending only on the shape of the spectrum
   (here $C = 0.637$);
λ is the mean wavelength;
$Q = \lambda/\Delta\lambda$ is a measure of the quality of the optical resonator and c is the velocity of light.

The coherence length Δl is written in the same notation:

$$\Delta l = c\,\Delta t = CQ\lambda.$$

By way of examples, holding to the same assumptions as above, we are able to quote white light for which Δl equals several microns, the yellow line of sodium for which $\lambda = 5890$ Å and $\Delta l \simeq 0.1$ Å corresponding to $Q \simeq 6 \cdot 10^4$ and $\Delta l = 2$ cm, the Hg^{198} line with $\lambda = 5461$ A for which $\Delta\lambda \simeq 10^{-2}$ Å and $\Delta l = 20$ cm, and finally a He and Ne laser for which $\lambda = 6328$ Å and $\Delta f = 10$khz. As $\Delta\lambda/\lambda = \Delta f/f$, $\Delta\lambda = \lambda^2\Delta f/c$, $Q = c/\lambda\Delta f$ and $\Delta l = Cc/\Delta f$, that is $\Delta l = 2 \cdot 10^6$ cm. The light is coherent over twenty kilometers. The number of periods is:

$$CQ = Cc/\lambda\Delta f = 3 \cdot 10^{10}.$$

Problem 17 (p. 19)

Let the mass m with abscissa x at the end of its spring be submitted to the force $F_0 \sin(\omega_0 t + \varphi_0)$.

We have the equation: $m\ddot{x} = -Kx + F_0 \sin(\omega_0 t + \varphi_0)$,

where

$$\ddot{x} = \frac{d^2x}{dt^2} = \frac{d}{dt}\left(\frac{dx}{dt}\right) = \frac{dv}{dt} \quad \text{with } x = \int v\,dt\,;$$

then

$$m\frac{dv}{dt} + K\int v\,dt = F_0 \sin(\omega_0 t + \varphi_0). \tag{1}$$

As the system is linear and because only a vibrating force with angular frequency ω_0 is applied, it will be necessary to consider only the value $\omega = \omega_0$ in the transformation of (1), since the system can only accept this value.

We transform equation (1);

$jm\omega V + \dfrac{K}{j\omega} V$, the transform of the first member is equal to the transform

of $F_0 \sin(\omega_0 t + \varphi_0)$. It follows that V, the transform of $v = dx/dt$, is equal to the product of the transform of the sine with $j\omega/(K - m\omega^2)$, that is, if we make $\omega = \omega_0$, with a factor equal to $j\omega_0/(K - m\omega_0^2)$.

It is known that $j\omega$ as a multiplying factor of a Fourier transform corresponds to a differentiation in the time domain. The velocity v will then be equal to:

$$F_0 \left[1/(K - m\omega^2) \right] \frac{d}{dt} \left(\sin(\omega_0 t + \varphi_0) \right)$$

$$v(t) = \left[\omega_0 F_0/(K - m\omega_0^2) \right] \cos(\omega_0 t + \varphi_0).$$

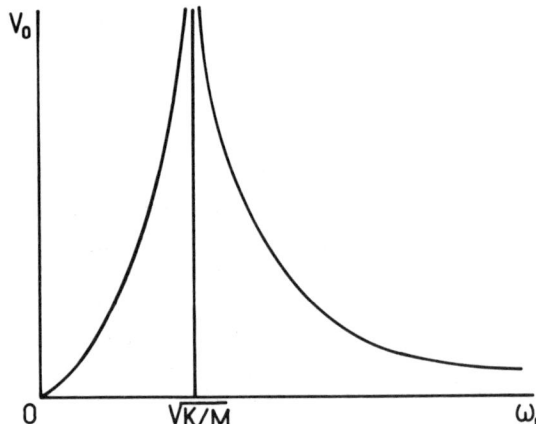

Figure 17.   Velocity of a mass m suspended by a spring whose spring constant is K and activated by a force $F_0 \sin(\omega_0 t + \varphi_0)$, as a function of ω_0.

$v(t)$ varies sinusoidally between $\dfrac{-F_0 \omega_0}{K - m\omega_0^2} = -v_0$ and $\dfrac{F_0 \omega_0}{K - m\omega_0^2} = v_0$.

v_0 tends to $+\infty$ as $\omega_0 \rightarrow \sqrt{K/m}$ (Figure 17).

There is resonance then without damping.

This physical reasoning is capable of being made more rigorous by the use of delta functions, as we shall be able to see later on. It will be noted that the solution of this problem by Fourier transformation is not particularly elegant. The method is more effective in the general case (see the following problem).

Problem 18 (p. 19)

When the coil circuit is open, there is no term of electromechanical coupling in the equations since there is no current passing through the coil.

The diagram of the mechanical part is shown in Figure 12. Let x be the abscissa of the coil measured on a vertical axis and relative to a fixed origin. The displacement u of the casing will be relative to the same axis and in the same direction. The equation of coil motion is written, with the same notation as in the previous assignment, in the form:

$$m\ddot{x} + K(x - u) = 0.$$

Now the voltage e at the coil terminals (circuit open) is proportional to the velocity of the coil relative to the casing:

$$e = K'\frac{d}{dt}(x - u).$$

If $X(f)$, $U(f)$, and $E(f)$ are the Fourier transforms of $x(t)$, $u(t)$, and $e(t)$, the two previous equations are transformed to give the system:

$$\begin{cases} -m\omega^2 X + K(X - U) = 0, \\ E = j\omega K'(X - U). \end{cases}$$

By elimination of X we deduce:

$$E = \frac{j\,m\omega^3\,K'\,U}{K - m\,\omega^2}.$$

With a harmonic excitation, the moving coil seismometer never supplies a voltage proportional to the displacement. At high frequency, the voltage is proportional to the ground velocity, for E is then equivalent to $-j\omega K'\,U$. When the frequency is very low, E is equivalent to $jm\omega^3\,(K'/K)\,U$, that is, the voltage is proportional to the derivative of the ground acceleration.

In the case where $u(t)$ is arbitrary, the voltage is obtained through a Fourier integral:

$$e(t) = \int_{-\infty}^{\infty} \frac{jm\omega^3 K' U}{K - m\omega^2}\, e^{j\omega t}\, df.$$

Generally, this integral is divergent since $\omega^3 U/(K - m\omega^2)$ tends to infinity as the numerator goes to zero. In practice, however, the voltage is limited by a damping which introduces a $j\omega$ term in the denominator, thereby preventing the numerator from ever going to zero. Furthermore, e cannot take very high values without the velocity, and also therefore the coil displacement, becoming very large. If such is the case, the coil then leaves the gap of the magnet and the corresponding nonlinearity prevents e from becoming enormous.

More details on seismometers and their filter properties will be found in later chapters.

Problem 19 (p. 19)

The equations capable of being handled by operational calculus are those which are linear, that is, the third and fifth equation. The first can also be considered as linear, but $1/y$ must be taken as a variable.

Problem 20 (p. 19)

$$S(f) = \int_{-\infty}^{\infty} s(t) \, e^{-j\omega t} \, dt.$$

If we differentiate with respect to f, we obtain

$$\frac{dS}{df} = -2\pi j \int_{-\infty}^{\infty} t \, s(t) \, e^{-j\omega t} \, dt.$$

It follows that the transform of $t \, s(t)$, if it exists, is $(j/2\pi) \, dS/df$.

If we now set $g = (e^{-\alpha^2 t^2})/t$, we see that the transform of tg is $(j/2\pi) \, dG/df$. Now tg has the simple form $e^{-\alpha^2 t^2}$ whose transform is $(\sqrt{\pi}/|\alpha|) \, e^{-(\pi f/\alpha)^2}$.

We therefore have

$$dG/df = -2\pi j \, (\sqrt{\pi}/|\alpha|) \, e^{-(\pi f/\alpha)^2}.$$

The spectrum sought is of the Gaussian function of the right-hand member in the last equation. Since g is odd and real, G is imaginary and odd. The integral selected will therefore be such that $G(0) = 0$.

Problem 21 (p. 19)

We write that $u(t)$ is the sum of 1/2 plus the function $u'(t)$ which has the value $-1/2$ for $t < 0$ and $+1/2$ for $t > 0$. The spectrum U of u is then the sum of $1/2 \, \delta \, (f)$ and the spectrum of u'.

But the derivative of u' is everywhere zero, except perhaps for $t = 0$. In fact, we are tempted to assign to it an infinite value at this position, which brings to mind the definition of $\delta \, (t)$. In actual fact, by considering that u' is the limit of a smoother function where the transition between the two levels is accomplished by a straight slope, we can follow a reasoning very close to that carried out to introduce the δ function. We see then that the integral of δ obtained is equal to the step of the function, that is, 1. It follows that $du'/dt = \delta \, (t)$.

The spectrum of the derivative is $j\omega U'$, U' being the spectrum of u'. Therefore $j\omega U' = 1$.

Finally

$$U = \frac{1}{2} \, \delta \, (f) + \frac{1}{j\omega}.$$

We may wonder why the derivative of u was not considered instead of taking that of u' directly. In actual fact $du/dt = \delta \, (t)$ as well, but it is obvious that the spectrum of u cannot be equal to $1/j\omega$. This function, being entirely imaginary, should in fact be the spectrum of an odd function. But u is not. The difficulty stems from the fact that the derivative defines a function only to within a constant. The correct value of this constant is the one that allows us to honor the theorems on Fourier transforms. We shall be more at ease in handling this kind of problem when generalized functions have been studied in detail. It will be noticed that Jennison (1961, p. 71) gives an incomplete result.

Problem 22 (p. 19)

Let $g(t)$ be real and symmetrical. Its transform $G(f)$, if it exists, is likewise real and symmetrical. It follows that the transform $H = j\omega G$ of $h = dg/dt$ is asymetrical.

$$- H(-f) = H(f).$$

Hence,
$$- h(-t) = h(t).$$

It is obvious that there is a simpler and more general explanation which allows us to bypass the Fourier transform.

Problem 23 (p. 20)

If we differentiate the triangle, we find two square pulses:

$$1 \text{ for } -T<t<o,$$
$$-1 \text{ for } 0<t<T.$$

The Fourier transform of the sum of these two square pulses is:

$$T (\text{sinc} fT) (- e^{-j\omega T/2} + e^{j\omega T/2}),$$

that is,

$$2jT \left(\sin \frac{\omega T}{2} \right) (\text{sinc} fT) = j\omega T^2 (\text{sinc} fT)^2.$$

The spectrum of the triangle is, therefore,

$$(T \text{ sinc} fT)^2.$$

The spectrum of $\delta(t+T) - 2\delta(t) + \delta(t-T)$, second derivative of the triangle, is equal to the product of the spectrum of this triangle with $(j\omega)^2$, that is $-4 \sin^2(\pi fT)$.

Problem 24 (p. 20)

If we assume that the variable density is correct, the variable density trace $h(t)$ is displayed as a darkening proportional to the actual trace $g(t)$, plus a constant a; the spectrum H of h is then equal to $a\delta(f) + G(f)$, G being the spectrum of g.

Since $h = a + g$, $dh/dt = dg/dt$. We see from this that two functions whose derivatives are equal have equal spectra except at the origin.

Problem 25 (p. 20)

Let the equation be

$$\int y \, dt + x + y(1-\alpha) = 0$$

We take the Fourier transform:

$$Y/j\omega + X + Y(1-\alpha) = 0.$$

That is,

$$(Y/X)\left[(1/j\omega) + 1 - \alpha\right] = -1.$$

Hence,

$$\frac{Y}{X} = -\frac{(1-\alpha)\,\omega^2 + j\omega}{1 + (1-\alpha)^2\,\omega^2}.$$

Let us examine $\dfrac{(1-\alpha)\,\omega^2}{1+(1-\alpha)^2\omega^2} + j\,\dfrac{\omega}{1+(1-\alpha)^2\omega^2}$ which represents $\dfrac{Y}{X}$ except for the sign.

We set $u = \dfrac{(1-\alpha)\,\omega^2}{[1+(1-\alpha)^2\omega^2]}$, the real part of $-Y/X$;

$v = \omega\,/\,[1+(1-\alpha)^2\omega^2]$, the imaginary part of $-Y/X$.

It is seen that $u^2 + v^2 = \omega^2\,/\,[1+(1-\alpha)^2\omega^2]$, which entails:

$$\left[u - \frac{1}{2(1-\alpha)}\right]^2 + v^2 = \frac{1}{4(1-\alpha)^2}.$$

We recognize here the equation of a family of circles (C) having their center on the u axis with the abscissa $1/2\,(1-\alpha)$ and having $1/2\,(1-\alpha)$ as radius. They are therefore tangent with the v axis at the origin.

For each value of a, the spectrum of $-Y/X$ is represented by one such circle in the plane defined by the real axis Ou and the imaginary axis Ov. We shall consider only the first quadrant of this plane.

If $\omega = 0$, we see that the phase is $\pi/2$ and the modulus is zero. When ω tends to ∞, the phase tends to 0 and the modulus tends to $1/(1-\alpha)$.

When $\alpha = 1$, the circle tends to the imaginary axis. The phase is equal to $\pi/2$ and the modulus is equal to ω.

On the other hand, it will be noted that:

$$1 - \alpha = \frac{1}{\omega}\frac{u}{v}.$$

Hence, by replacing $1-\alpha$ in the expression of v:

$$v = \frac{\omega}{1 + \dfrac{u^2}{v^2}},$$

which can be written:

$$u^2 - \omega v + v^2 = 0$$

or: $$u^2 + (v - \omega / 2)^2 = \omega^2/4.$$

This is the equation of a family of circles (Ω) centered on the imaginary axis with abscissa $\omega/2$, radius $\omega/2$, and tangent to the real axis at the origin.

In order to have the modulus and phase of $-Y/X$ for given values of ω and α, we take the intersection of the circle (C) corresponding to the value of α with the circle (Ω) corresponding to the value of ω (Figure 18).

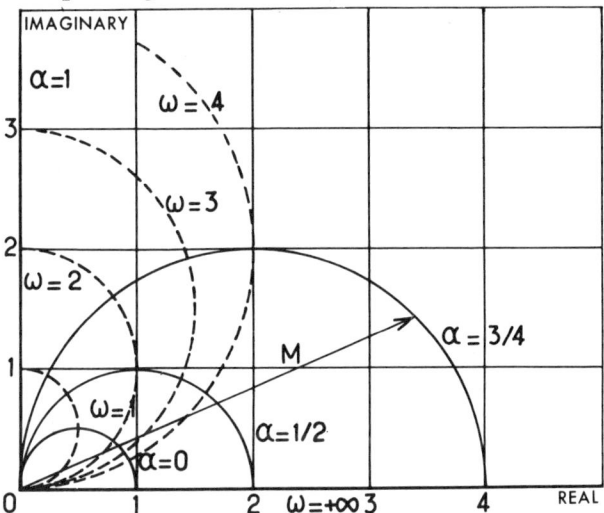

Figure 18.   Graph to determine the amplitude and phase of the ratio Y/X for given values of α and ω.

We can thus easily construct the representation of the modulus as a function of ω (Figure 19).

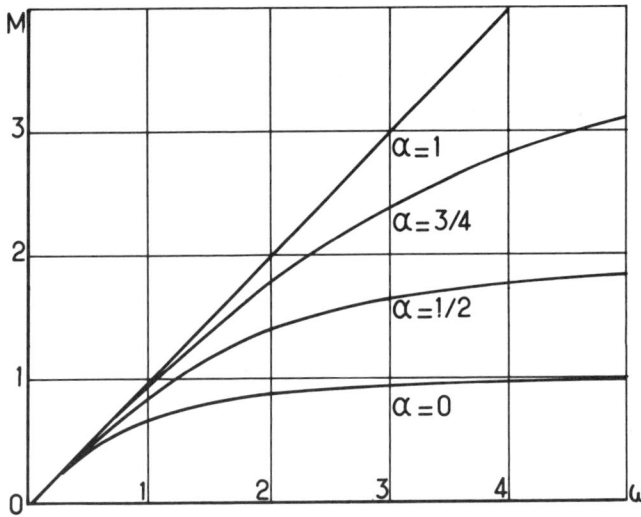

Figure 19.   Representation of the amplitude of the ratio Y/X as a function of ω.

For small values of ω, the ratio $-Y/X$ is equivalent to $j\omega$; $-y$ is then the derivative of x. For a given value of ω, this is especially true as α approaches 1. In fact, by expanding the denominator we have

$$- Y/X = [j\omega + (1-\alpha)\,\omega^2]\,[1 - (1-\alpha)^2\omega^2 + (1-\alpha)^4\omega^4 - ...].$$

Furthermore, it is to the high frequencies that the portion of the circle (C) close to the real axis corresponds. For high frequencies the modulus remains approximately constant and close to $1/(1-\alpha)$ and the phase approaches 0. The high frequencies are not amplified more than the medium frequencies. We therefore accomplish a differentiation at the low frequencies without excessively amplifying the high frequencies.

This is not true for $\alpha = 1$ because the amplification is then proportional to ω.

Problem 26 (p. 27)

The receiver is at a finite distance; we look for the diffraction pattern at infinity. We set

$$\begin{cases} G\,(R) = 1 & \text{for} \ \ R < R_0 \\ \quad\ = 0 & \text{for} \ \ R > R_0. \end{cases}$$

The function which gives the directivity is

$$g\,(r) = 2\pi \int_0^\infty R\,G\,(R)\,J_0\,(2\pi r R)\,dR$$

$$= 2\pi \int_0^{R_0} R\,J_0\,(2\pi r R)\,dR.$$

If we now make $n = 1$ in the derivative formula of the Bessel functions which is given in the text, we have

$$\frac{d}{du}\,[u\,J_1\,(u)] = u\,J_0\,(u).$$

This formula applies here if we set $u = 2\pi r R$, which gives

$$g\,(r) = \frac{1}{2\pi r^2} \int_0^{2\pi r R_0} u\,J_0\,(u)\,du$$

$$= \frac{1}{2\pi r^2}\,[u\,J_1\,(u)]_0^{2\pi r R_0}$$

$$= \frac{R_0}{r}\,J_1\,(2\pi r R_0).$$

By making $R_0 = 1$, we obtain

$$g(r) = \frac{1}{r} J_1(2\pi r).$$

We see that this formula is similar to the one giving the spectrum of a square pulse:

$$\text{Cren } (t/T) \longleftrightarrow |T| \text{ sinc } (fT),$$

by noting

$$\text{Cren } (t/T) = 1 \quad \text{for} \quad |t| < |T|/2$$
$$= 0 \quad \text{for} \quad |t| > |T|/2.$$

From the appearance of $\dfrac{J_1(\pi u)}{\pi u}$, it is seen that the diffraction pattern can be written in a form containing a Bessel function analogous to the sinc function.

Let us see if $g(0) = 2\pi \displaystyle\int\limits_0^\infty R\, G(R)\, dR.$

The integral equals

$$2\pi \int\limits_0^{R_0} R\, dR = \pi\, R_0^2.$$

Also

$$g(0) = R_0 \lim_{r \to 0} \left[(J_1(2\pi r R_0))/r \right] = R_0\, \pi\, R_0 = \pi R_0^2.$$

Problem 27 (p. 27)

The transform is radial and is written

$$2\pi \int\limits_0^\infty R\, \frac{1}{R} J_0(2\pi r R)\, dR,$$

that is, by setting $u = 2\pi r R$,

$$\int\limits_0^\infty J_0(u)\, \frac{du}{r}.$$

But it is known that the integral of J_0 equals 1. It follows that $1/R \longleftrightarrow 1/r$.

Problem 28 (p. 27)

Let $g = e^{-\pi r^2} = e^{-\pi(k_1^2 + k_2^2)}.$

$$G = \iint\limits_{-\infty}^{\infty} e^{-\pi(k_1^2 + k_2^2)}\, e^{-2\pi j(k_1 x_1 + k_2 x_2)}\, dk_1\, dk_2.$$

This integral separates into two parts:

$$G = \int_{-\infty}^{\infty} e^{-\pi k_1^2} e^{-2\pi j k_1 x_1} \, dk_1 \cdot \int_{-\infty}^{\infty} e^{-\pi k_2^2} e^{-2\pi j k_2 x_2} \, dk_2$$

$$= e^{-\pi x_1^2} \cdot e^{-\pi x_2^2} = e^{-\pi(x_1^2 + x_2^2)} = e^{-\pi R^2}.$$

We can verify the theorem which is explained directly below the statement of this problem. It is written

$$(e^{-\pi R^2})_{R=0} = 2\pi \int_0^{\infty} r e^{-\pi r^2} \, dr,$$

that is

$$1 = \int_0^{\infty} e^{-\pi r^2} \, 2\pi r \, dr = \int_0^{\infty} e^{-u} \, du$$

if $u = \pi r^2$.

Problem 29 (p. 31)

Let the function be $g(x_1, x_2)$ and its Fourier transform

$$G(k_1, k_2) = \int\!\!\!\int_{-\infty}^{\infty} g(x_1, x_2) e^{-2\pi j(k_1 x_1 + k_2 x_2)} \, dx_1 \, dx_2.$$

If we perform a rotation of the axes Ox_1 and Ox_2 through angle θ which leads to Ox_1' and Ox_2', $g(x_1, x_2)$ is transformed to $g(x_1', x_2')$, obtained by replacing x_1 and x_2 by their values as a function of x_1' and x'_2:

$$\begin{cases} x_1 = x_1' \cos \theta - x_2' \sin \theta, \\ x_2 = x_1' \sin \theta + x_2' \cos \theta. \end{cases}$$

If we are looking for the transform $G'(k_1, k_2)$ of $g(x_1', x_2')$, we must calculate

$$\int\!\!\!\int_{-\infty}^{\infty} g(x_1', x_2') e^{-2\pi j(k_1 x_1 + k_2 x_2)} \, dx_1 \, dx_2.$$

By changing the variable from x to x', we have

$$G'(k_1, k_2) = \int\!\!\!\int_{-\infty}^{\infty} g(x_1', x_2') e^{-2\pi j[k_1(x_1' \cos \theta - x_2' \sin \theta) + k_2(x_1' \sin \theta + x_2' \cos \theta)]} \, dx_1' \, dx_2'.$$

As a matter of fact, $dx_1 \, dx_2 = dx_1' \, dx_2'$ because the transformation matrix

$$\begin{pmatrix} \cos \theta & - \sin \theta \\ \sin \theta & \cos \theta \end{pmatrix}$$

has a determinant equal to 1.

If we redistribute the terms of the exponent so as to make k_1' and k'_2 appear as

$$\begin{cases} k_1' = k_1 \cos \theta + k_2 \sin \theta, \\ k_2' = - k_1 \sin \theta + k_2 \cos \theta, \end{cases}$$

the integral takes the form:

$$\int\!\!\!\int_{-\infty}^{\infty} g\,(x_1', x_2')\, e^{-2\pi j(k_1' x_1' + k' x_2')}\, dx'_1\, dx'_2.$$

It follows, from the written identity of this integral with the integral giving $G\,(k_1, k_2)$ from $g\,(x_1, x_2)$, that $G'\,(k_1, k_2) = G\,(k_1, k_2)$. The transformation which changes k to k' is exactly the same as that changing x to x', which demonstrates the theorem.

The reader will have no difficulty in confirming physically the exactness of this theorem by means of experimenting with optical diffraction.

Chapter 4

DELAY-LINE FILTER
OF THE FRENCH PETROLEUM INSTITUTE

B. DAMOTTE

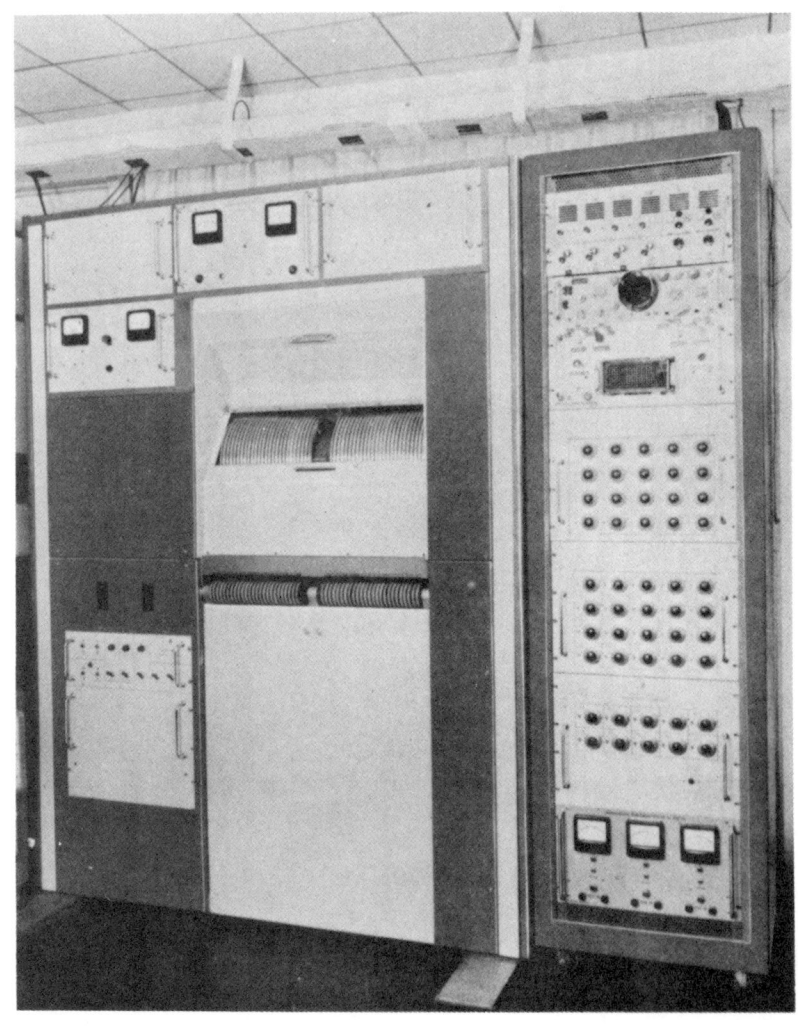

INTRODUCTION

Information gathered in the course of seismic exploration and stored on magnetic tape is made up of a mixture of signal and noise in varying proportions.

The aim of every geophysicist is to reconstitute records with an increased signal-to-noise ratio to simplify interpretation. During replay of the magnetic tape, selection is made of the useful signal from other less useful signals and from the noise. This selection is evaluated in the frequency domain. Recordings carried out in marine seismic work illustrate this phenomenon; in order for them to be usable, the spurious effect arising from resonance of the water layer must be removed.

For this purpose the geophysicist resorts to two main procedures at the analog playback center: electrical filtering in the frequency domain and delay-line filtering in the time domain. Unlike the former which allow him only passbands, the delay-line filters provide broader possibilities. This kind of equipment can, in addition to the passbands, achieve band-reject filters of different width and slopes going from suppression of a fixed frequency, as for example the 60-hz industrial power, to attenuation of a wide frequency band as is the case in deconvolution procedures. Narrow rejection bands can still be regularly spaced over the spectrum allowing for removal of multiples that may exist in land or marine seismic.

DEFINITIONS OF LINEAR FILTERING

Before presenting the delay-line filter of the French Petroleum Institute (IFP), it is appropriate to recall briefly linear filter theory and to see how, by operating in the time domain, filtering can be carried out with the delay-line filter.

It is known that a linear filter can be defined in two equivalent ways: in the frequency domain by its transfer function or in the time domain by its impulse response.

Frequency domain

A seismic trace may be regarded as being the sum of a certain number of sinusoids, of different frequency, each sinusoid having its own amplitude and phase. It can therefore be expressed as a function of time (Figure 20A) by the Fourier integral:

$$k(t) = \int_{-\infty}^{+\infty} p_k(f) \cos[2\pi ft + \varphi_k(f)] \, df$$

or as a function of frequency by its spectrum:

$$K(f) = p_k(f)e^{j\varphi_k(f)}.$$

These are expressions in which $p_k(f)$ and $\varphi_k(f)$ are the amplitude and phase for each sinusoid of frequency f.

The filter action is going to consist of changing the amplitude and phase of certain sinusoids in the seismic trace. This change will be made in accordance with the amplitude response and phase characteristic of the filter considered (Figure 20B), thereby transforming the spectrum $K(f)$ of the original trace into a new spectrum $G(f)$ which is that of the filtered trace. The operation is written as:

$$G(f) = K(f)\,T(f),$$

where $T(f) = p(f)e^{j\varphi(f)}$ is the transfer function characteristic of the filter.

The filtered trace has the spectrum:

$$G(f) = p_G(f)e^{j\varphi_G(f)}$$

and can also be expressed in the form of the Fourier integral:

$$g(t) = \int_{-\infty}^{+\infty} p_G(f)\,\cos\,[2\,\pi f t + \varphi_G(f)]\,df \quad \text{(Figure 20C)}.$$

Time Domain

The filter is characterized in the time domain by its response to a unit impulse. This response is called the impulse response $R(\tau)$ or the filter operator (Figure 20E). A seismic trace $k(t)$ may be considered as a time series of impulses with varying amplitude and sign (Figure 20D), in which the sampling interval is infinitely small.

Each basic impulse $k(t_n)$ gives an impulse response in passing through the filter. The impulse response is proportional in amplitude and sign to $k(t_n)$. The operation is written as:

$$g_n(t) = k(t_n)\,R(\tau).$$

A series of impulse responses of varying amplitude and sign, overlapping in time, is obtained in this manner (Figure 20F). The ordinate of the filtered trace at time t_n is the algebraic sum of the ordinates for the overlapping impulse responses; in other words:

$$g(t_n) = \Sigma\,g_n(t) = \underset{n}{\Sigma}\,k(t_n)\,R(\tau),$$

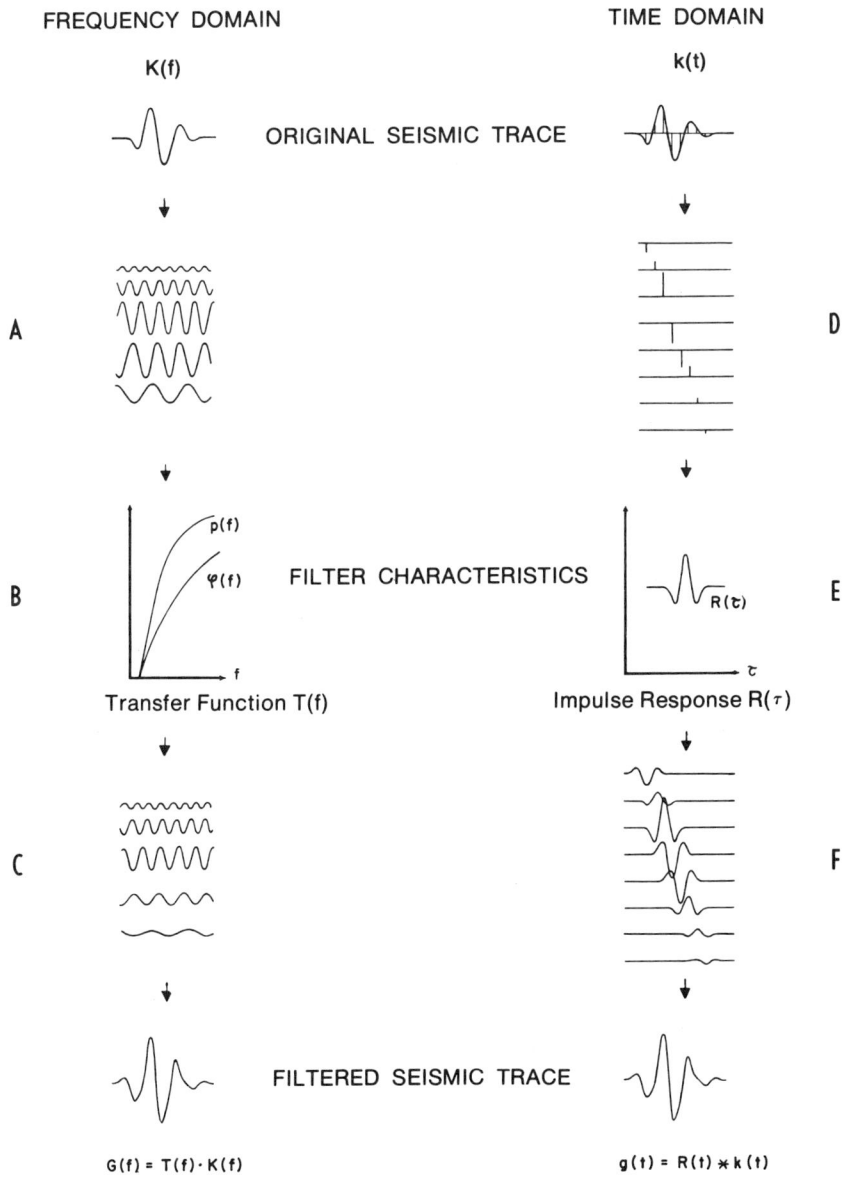

FREQUENCY DOMAIN

TIME DOMAIN

K(f)

k(t)

ORIGINAL SEISMIC TRACE

A

D

B

FILTER CHARACTERISTICS

E

Transfer Function T(f)

Impulse Response R(τ)

C

F

FILTERED SEISMIC TRACE

G(f) = T(f)·K(f)

g(t) = R(t) ∗ k(t)

Figure 20. Flow diagram of linear filtering.

and the filtered trace is expressed by the convolution integral:

$$g\,(t) = \int_0^t k\,(t - \tau)\, R\,(\tau)d\tau$$

by setting $t = t_n + \tau.$

This result can again be expressed as follows: The filtered trace is equal to the original trace convolved with the filter operator

$$g\,(t) = k\,(t) * R\,(t).$$

These two definitions of linear filtering are equivalent; this equivalence between the transfer function $T(f)$ and the impulse response $R\,(\tau)$ is given by the Fourier transforms:

$$R\,(\tau) = \int_{-\infty}^{+\infty} T\,(f) e^{2\pi i f \tau}\ df,$$

$$T\,(f) = \int_{-\infty}^{+\infty} R\,(\tau) e^{-2\pi i f \tau}\ d\tau.$$

PRINCIPLE OF DELAY-LINE FILTERING

The delay-line filter is an analog device which implements filters characterized by their impulse response.

It resolves the equation:

$$g\,(t) = \int_{-\infty}^{+\infty} k\,(t - \tau)\,R\,(\tau)\,d\tau,$$

or the convolution product:

$$g\,(t) = k\,(t) * R\,(t),$$

where $g(t)$ is the filtered trace; $k(t)$ is the original trace; $R(\tau)$ is the filter impulse response or filter operator.

In order to understand better the mechanism of linear filtering in the time domain, the seismic trace $k(t)$ to be filtered has previously been considered as a suite of samples, and the impulse response $R\,(\tau)$ as a continuous function. We can also proceed in the inverse manner to obtain the same result, that is to find the filtered trace $g(t)$ (always provided that the sampling of R is correct). It is in the latter sense that the delay-line filter works effectively.

On the assumption that the sampling capacity of such equipment is strictly limited, it is preferable to sample the operator. A delay line therefore implements filtering according to the following principle (Figure 21A).

The seismic trace $k(t)$ is transcribed in continuous fashion on a magnetic drum rotating at constant speed V, then read again with progressively larger delays through N heads regularly spaced around the drum. The trace $k(t)$ is thus read N times, each reading being delayed by a time τ with respect to the previous

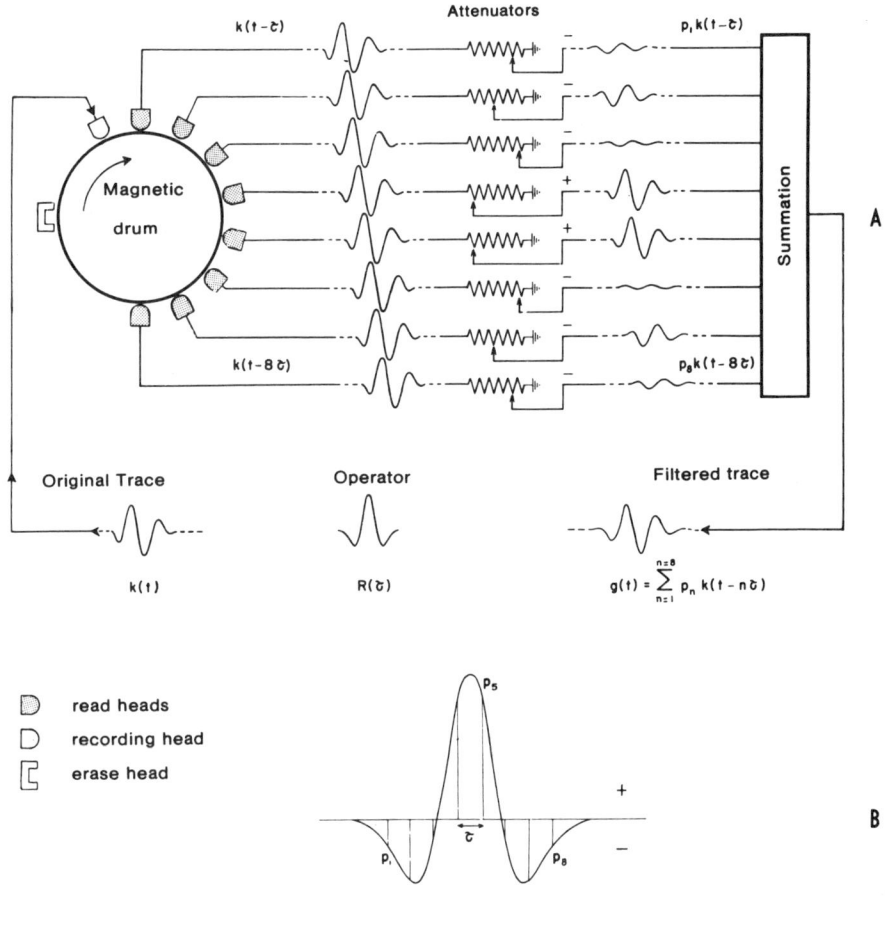

Figure 21. Principle of a delay-line filter.

one. This interval τ depends on the speed V of the drum and the separation between each read head. Each of the N traces is then subjected to a weighting p and occasionally a change of polarity. Finally, these N traces delayed relative to each other, weighted and with appropriate sign, pass through a summing circuit, from which the filtered trace $g(t)$ is output.

The operation is written as:

$$g(t) = \sum_{N} p_N k(t - N\tau),$$

where the operator is introduced in the form of N samples regularly spaced by τ msec and whose ordinates are proportional in value and sign to the weighting p_N (Figure 21B).

IFP DELAY-LINE FILTER

General

The delay-line filter constructed at the French Petroleum Institute and displayed in Figure 23 conforms to the previously expressed principle, with the difference that instead of having only a single recording head for a given number of read heads it provides for an equal number of heads of each type.

The number of pairs is 50. The read heads are fixed and arranged according to the magnetic drum generator. The corresponding recording heads, each fixed to a ring completely independent of the other, can be moved around the edge of the drum so as to introduce variable delays.

The diagram in Figure 22 shows the drum and its concentric rings. The front view is identical to that seen on the photograph of the equipment (Figure 23) where part of the drum can be discerned between two groups of 25 rings.

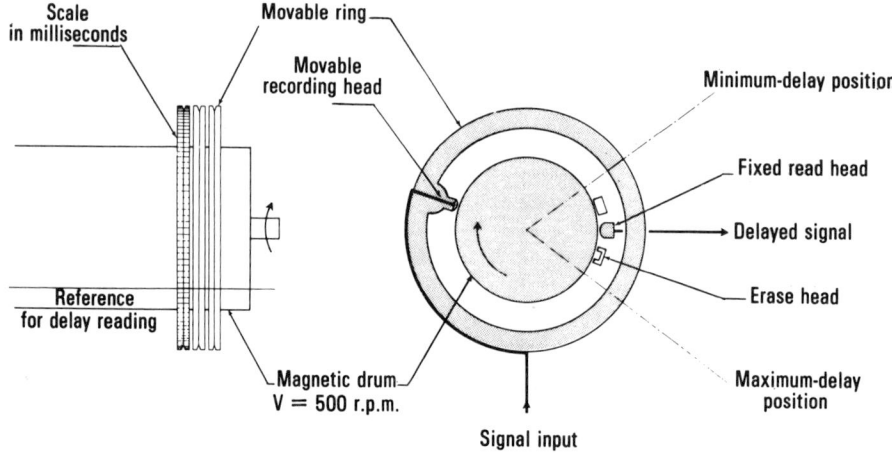

Figure 22.   Diagram of magnetic drum and heads.

The side view shows the arrangement of each pair of heads. The recording head forms part of the movable ring which is graduated in milliseconds of delay with respect to the fixed read head. Maximum delay is obtained by positioning the recording head at the side of the erase-head, and minimum delay by positioning in a symmetrical manner relative to the read-head.

Unlike the first procedure with fixed heads which permits only a single value of sampling interval τ, if the rotation speed of the drum is constant, the procedure employed here provides wide flexibility of use; τ may, in fact, be variable or even irregular. Furthermore, the point in having as many recording heads as read heads allows the equipment to be used for filters which require input of several different

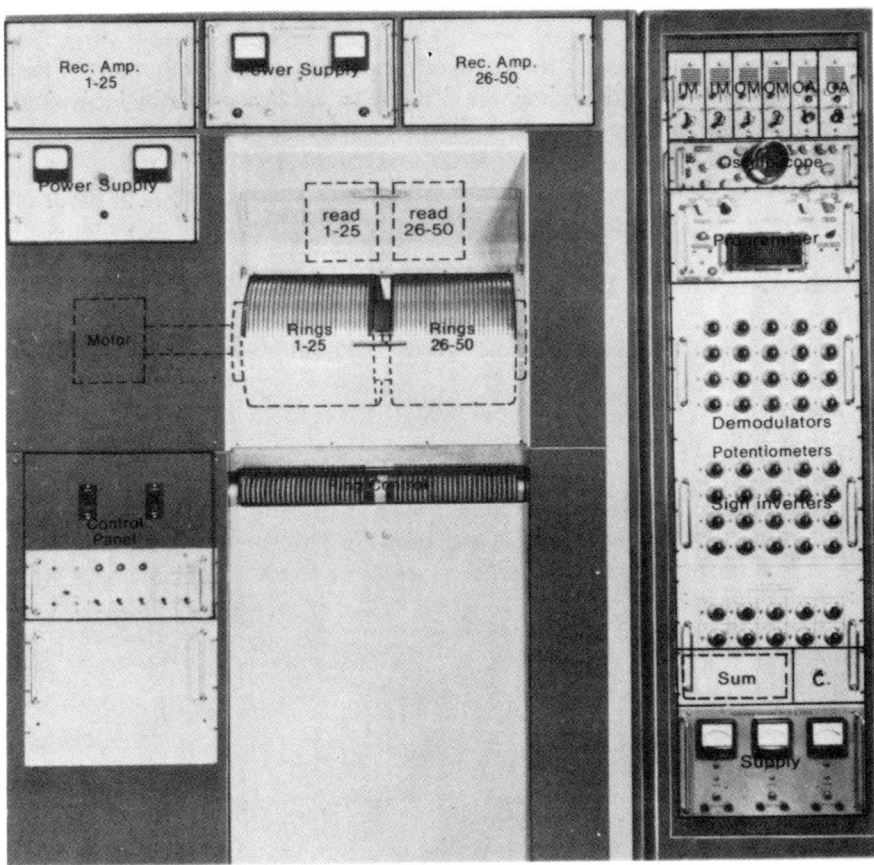

Rec. Amp.:	Recording amplifiers	O.A.	Output amplifiers
Pl. Preamp.:	Read preamplifiers	Sum.:	Summation resistances
I.M.:	Input modulators	S.:	Switch for in parallel arrangement
			of channels 1 to 25 and 26 to 50.
O.M.	Output modulators	Supply:	Power supply

Figure 23.   Arrangement of delay-line filter components.

traces (stacking, fan filter) or for performing simultaneous static corrections with groups of 24 seismic traces.

Description of delay-line filter

As mentioned before, the IFP equipment comprises 50 channels. This number is therefore the maximum number of samples available to display an operator.

Electronics

The equipment operates by frequency modulation. The frequency of the carrier wave is 4000 hz and the deviation is ± 200 hz. A modulator located at the input permits the introduction of low-frequency signals, on occasion.

Since the IFP delay-line filter is an arrangement of 50 recording heads, the seismic trace $k(t)$ must be applied in parallel. This is the function of the program circuit board. This component constitutes the turntable of the machine. Its main role is to feed the modulated signals over the required number of channels to accomplish the desired operation, by means of connections plugged into a detachable board.

Each of these 50 channels consist of the following components (see the work diagram, Figure 24):

    — recording amplifier, with tubes, which drives the heads at an adequate level;

    — recording head;

    — drum;

    — read head. The very weak signal read is preamplified and reshaped by a transistor preamplifier located very close to the heads in order to avoid the introduction of any extraneous noise to the signal during its transfer at low level;

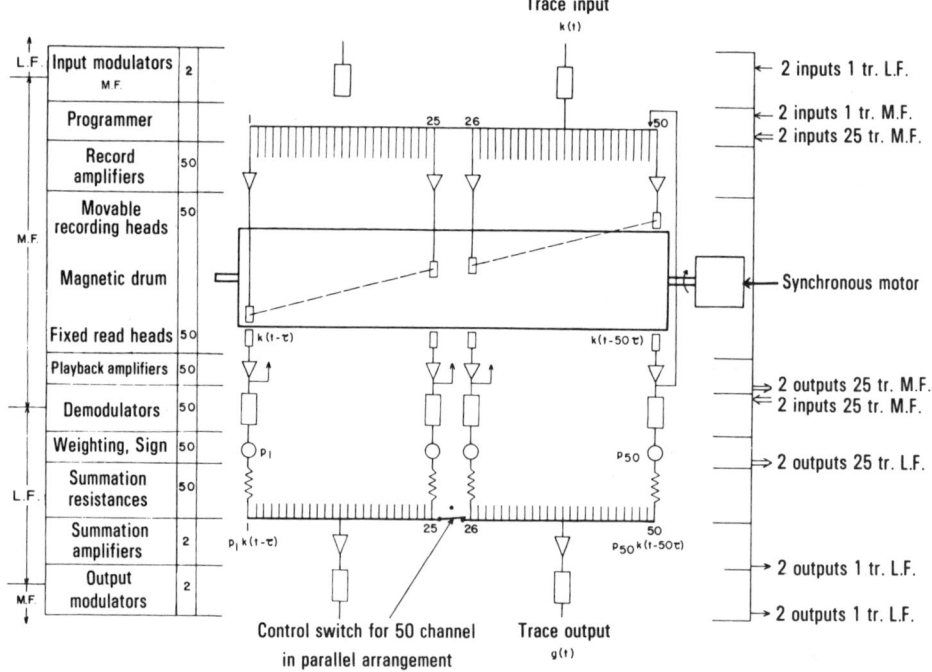

Figure 24.   Schematic diagram of delay-line filter.

— transistor demodulator which restores the seismic frequencies. An attenuator and a polarity inverter act on this low frequency to apply the weighting and sign of the corresponding operator sample.

— a summing resistance completes the individual circuit of each channel.

The output of the 50 channels, that is the 50 original seismic traces displayed relative to each other, weighted and treated for sign, is then summed.

This unmodulated trace passes through a summing tube amplifier to make up the filtered trace. Next, according to whether we wish to record on magnetic tape or photograph, it is made to pass through a frequency modulator or bypassed.

To enhance the possibilities of the equipment we can split the 50-channel delay-line filter into two independent filters each with 25 channels. As a result, an additional capability exists for input modulators, summation amplifiers, and output modulators.

As shown in the diagram of Figure 24, the equipment incorporates the following components in the order that the seismic trace goes through them:

— 2 input modulators;
— program circuit board;
— 50 recording amplifiers;
— 50 movable magnetic recording heads;
— magnetic drum;
— 50 fixed magnetic read heads;
— 50 playback preamplifiers;
— 50 frequency demodulators;
— 50 attenuators and polarity inverters;
— 50 summation resistances;
— 2 summation amplifiers;
— 2 output modulators.

In order to increase the flexibility of the equipment in its different applications, provision has been made for a number of inputs and outputs which are listed below (see also the diagram in Figure 24):

a) Inputs and outputs for a single trace (sequential time filter).

The trace $k(t)$ to be filtered may be input:
— either in low frequency on the input modulator;
— or frequency modulated on the program circuit board.

The filtered trace $g(t)$ may be output:
— either in low frequency from the summation amplifier;
— or frequency modulated from the output modulator.

b) Inputs and outputs for 25 traces (spatial filtering, static corrections):

25 different traces frequency modulated may be input simultaneously:
— either on the program circuit board;
— or at the input of the demodulators.

25 different traces may be output simultaneously:
— either frequency modulated from the playback preamplifiers;

— or in low frequency from the demodulators[3].

The modulated output of the playback preamplifiers is delivered by the program circuit board. The signal produced by these preamplifiers is merely delayed; by reintroducing it on another channel, the output delay capacity of the equipment is increased. This innovation enables us to apply operators whose duration exceeds 96 msec, or further, to static correct a trace with a maximum time of 5388 msec by employing the 50 channels.

To change the 50-channel delay-line filter into two filters with 25 channels, beyond the program requirements of this modification, it is necessary to actuate a switch with two positions, one of which makes a parallel connection of the outputs of channels 1 to 25 and another one of those of channels 26 to 50. In case of the 50-channel filter, all corresponding outputs are in parallel.

Mechanics

The delay that may be introduced in each channel ranges from 12 to 108 msec. This minimum of 12 msec is imposed because we cannot physically position the recording heads on top of the read heads (Figure 22) to obtain a zero delay. The 12 msec therefore corresponds to the portion of the circumference rendered unusable by the crowding of the heads and by the erase-head in the extreme positions.

Each movable ring carrying a recording head is graduated in milliseconds. Since the portion of the ring circumference corresponding to one millisecond is somewhat more than 8 mm, we can easily adjust to a quarter of a millisecond.

The drum coated with a base of magnetic oxide is driven by a synchronous three-phase motor connected to the mains, with a speed of 500 rpm. This speed is permanently controlled by an electronic counter. A small electromagnet attached very close to a notched wheel, located at the end of the drum axle, sends an impulse to the counter with the passage of each notch.

The weight of the drum which is 40 kg, the supporting frame cast of Niresist (nonmagnetic) metal and weighing more than 250 kg, the nonrigid motor coupling and its attachment ensure that no vibration is felt at the level of the magnetic heads.

The magnetic heads are located 0.03 mm (± 0.01) away from the surface of the drum. We thereby avoid all signal variations which, in the case of heads rubbing, are produced by wear of the heads or of the magnetic material.

Performance and characteristics

Delay

We have seen earlier that a delay of 96 msec is introduced on each ring, but that more significant delays are possible by looping a previously delayed signal through another channel.

---

[3]This description relates to the use of a single delay-line filter (25 or 50 channels). The different inputs and outputs are in duplicate for use as two separate delay-line filters.

In time-domain filtering, an operator may be represented by 50 samples whose sampling interval τ may be as small as 0.25 msec, and as large as 96 msec.

For simultaneous static correction with 25 traces, the maximum delay is $96 + 108 = 204$ msec.

For sequential static correction (refraction) the maximum delay is 5.388 sec.

Weighting

Each potentiometer is graduated in units from 0 to 500.

Passband

The circuits which make up the equipment have been studied for the for the $0.25 - 400$ hz passband. These values are intended for a 3 db attenuation.

Dynamic range

The dynamic range reaches 53 db when measured on a single channel; for 50 channels it is 55 db. To evaluate this dynamic range, we introduce a sinusoid of seismic frequency to the modulator input. We measure the output level of the summation amplifier with identical delay, weighting and sign on each channel, first of all by modulating 100 percent to obtain maximum signal, then bringing down the modulation to obtain the noise level.

Power requirements

The equipment is connected to 127v 50-cycle mains and uses 15 amps. The motor is independently supplied by triphased 220v 50-cycle power and its consumption is 3 amps.

Supplies from the mains that can be seen in the photograph (Figure 23) provide the required continuous voltages:

300v located in the upper central part of the equipment for the record amplifiers and modulators;

12v A for the playback preamplifiers and the demodulators;

12v B and -21v for the demodulators;

These last three supplies are located at the bottom of the instrument panel;

12.6v located on the left side of the equipment below the recording amplifiers, for heating the modulator tubes.

Heating of the recording amplifier tubes is accomplished by alternating current through transformers connected to the mains.

Current for the summing amplifiers is supplied directly from the mains.

Programming

Description of program circuit board (Figure 25)

Use of the delay-line filter as a sequential filter, spatial filter or static corrector requires special programming for each case.

The fixed part of the program circuit board is made up of a small plate with 240 sockets, each one of which has a corresponding contact at the base. Connector leads allow us to join two of these contacts by plugging their ends into the corresponding sockets. The sockets are arranged in 10 lines of 24 and are partitioned in this way.

The 50-socket group in the upper section (lines 1 and 2) facilitates input of 50 frequency modulated traces into the equipment. They are directly linked to two inputs of 25 FM traces in the work diagram (Figure 24). A certain number of them are used in spatial filtering and static correction.

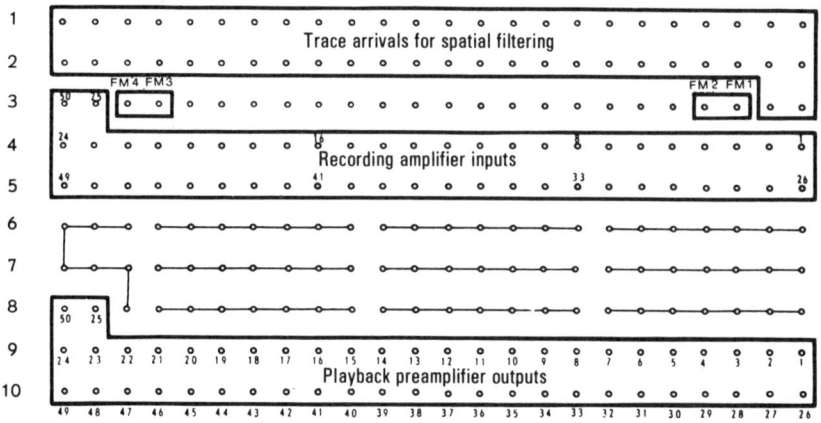

Figure 25.   Diagram of programming circuit board.

The 50 sockets distributed mainly on lines 4 and 5 are the inputs of 50 channels. Their corresponding contact is connected to the input of the recording amplifiers.

The sockets marked FM1, FM2, FM3, FM4 on both sides of the third line correspond to the output of four frequency modulators (two input, FM1 and FM2, two output, FM3 and FM4). One of the sockets accommodates the arrival of the signal which, in time-domain filtering, must be transmitted in parallel along the different channels. The other sockets on the third line are unused.

Sockets of lines 6, 7, and 8 are interconnected in groups of seven, as shown in the diagram. They serve to effect sending the signal in parallel over the 50 channels.

The 50-socket group in the lower section (lines 9 and 10) is in contact with the playback preamplifier outputs. A modulated signal is found here, delayed by the value set on the corresponding ring, and amplified to the same level as

the signal leaving a modulator. This allows us to reintroduce the signal into another channel in order to increase the delay.

Different types of programs

Time-domain filtering — The original trace $k(t)$ has to be sent in-parallel over
the number of channels required to represent the operator.

α) *operator has a duration less than 96 msec (Figure 26A).*

The trace arrives modulated at "MF1" and through the medium of lines 6, 7,
and 8, is sent simultaneously over the requisite channels selected on lines 3, 4,
and 5.

β) *operator has a duration greater than 96 msec (Figure 26B).*

The trace arrives modulated at "MF1" and as long as the delay of 96 msec
is not reached, we proceed as above. For larger delays, the signal that must be
sent in-parallel through the corresponding channels is taken in one of the sockets
of lines 8 (two left-hand sockets), 9 and 10 (outputs of playback preamplifiers).
These latter channels therefore possess, over and above their own delay, a common initial delay which is that of the supply channel selected.

As can be seen in the work diagram (Figure 24), the action of reintroducing
a previously delayed trace into another group of channels in no way prevents this
same trace from participating in the final summation.

In the choice of channel subjected to the additional delay, the 12 msec mechanical gap must also be taken into account. In other words, the first delay of
a new series should be equal to the desired delay minus the delay of the carrier
channel. The difference between these two delays should be equal to or greater
than 12 msec.

γ) *example*

A 50-term operator with $τ = 3$ msec lasts 147 msec. We must therefore set
the delays $147 + 12 = 159$ msec, distributed from 12 to 159. We reach 108 msec|
on ring 33. On account of the 12-msec gap, we introduce the signal produced
from channel 30 which is delayed by 99 msec. Channels 34 to 50 consequently
have an initial delay of 99 msec. We then set a delay of 12 msec ($+99 = 111$)
on ring 34 and a delay of 60 msec ($+99 = 159$) on ring 50.

Spatial filtering (Figure 26C) — The traces to be added appear frequency
modulated on one of the first two lines of the board and each of them must be
connected to a recording amplifier from lines 4 and 5. According to whether the
addition is oblique or straight, delays are introduced or omitted on the ring. If
the addition is not weighted, the potentiometers of channels used must be adjusted
to the same level; in the opposite case, they are adjusted to the weighting.

Static correction (Figure 26C) — The program is the same as before. The delays
should correspond to the desired corrections. We can output the corrected traces,
either modulated at the outlet of the playback preamplifiers, or after demodulation at the outlet of the demodulators (the sign or gain of certain traces can

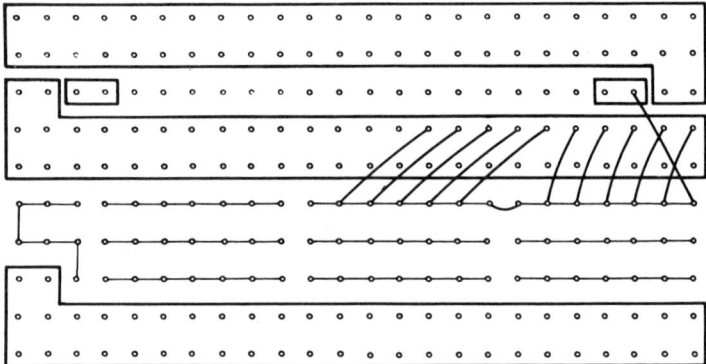

A. Time-domain filtering. Operator less than 96 ms.

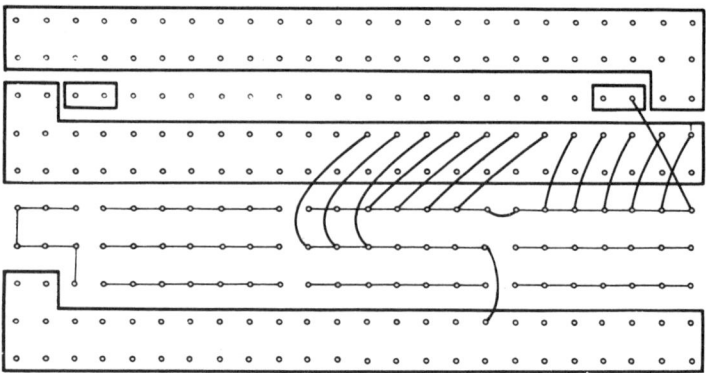

B. Time-domain filtering. Operator greater than 96 ms.

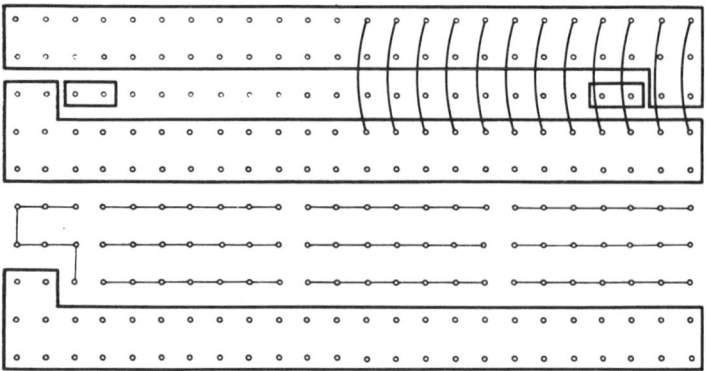

C. Spatial filtering and static corrections.

Figure 26.   Different types of programs.

thus be changed on occasion).

Processing facilities and examples

As has already been suggested in the preceding paragraphs, the facilities offered by the IFP delay-line filter are manifold.

Single machine use

Time-domain filtering — This filtering is carried out sequentially and concerns therefore only a single trace at a time. It is solely this kind of filtering that can be implemented on a machine comprising only a single recording head and several read heads.

This filtering is employed to cut out harmful frequencies in the seismic band, notably filtering of the 50-hz industrial power current or the reverberation effect in marine seismic. Setting out from the same principle, it can also remove multiple reflections.

It is further utilized to attenuate certain dominant frequencies in the spectrum of a seismic trace in order to ensure that each component frequency of this trace has an equivalent contribution. This is the deconvolution process for which it is possible to locate the desired attenuation at a particular place in the spectrum.

Figures 27a and 27b show some examples of filtering in marine seismic (A-B), filtering multiples in land seismic (C), filtering of 50 hz (D), and deconvolution (E).

Spatial filtering — It is necessary here to introduce several traces simultaneously into the machine. They need not be consecutive traces. In this way we implement the entire range of trace addition and mixing, with or without weighting, with oblique or straight addition.

Time-space filtering — This type of filtering acts on the apparent velocities of seismic data without attenuation of the high frequencies. It consists of performing spatial filtering on a certain number of traces (most often 12) which have previously been frequency filtered individually in a different way.

The example in Figure 28 shows a record processed by a trapezoidal operator (Embree et al, 1963; Fail and Grau, 1963; Fail et al, 1964). Each of the 12 original traces is sent through a time-domain filter F_n effected by four channels. These F_n filters are of six different types, symmetrically arranged as shown in the diagram in the upper part of Figure 28. Each of them is characterized by an impulse response or four-term operator in which the sampling interval τ, as well as the weighting, is variable (Figure 28, $R(\tau)$ of F_1 and F_6).

The outputs of the 12 groups of 4 channels are summed to achieve the spatial filter and thereby furnish the "fan-filtered" trace.

The program combines the two basic programs of spatial filtering and time-domain filtering (programming diagram, Figure 28). The traces arrive modulated in frequency at the first 12 outlets of the first line and each of them is sent in-parallel on four channels (trace 1 on channels 1 to 4, trace 2 on channels 5 to 8, etc.)

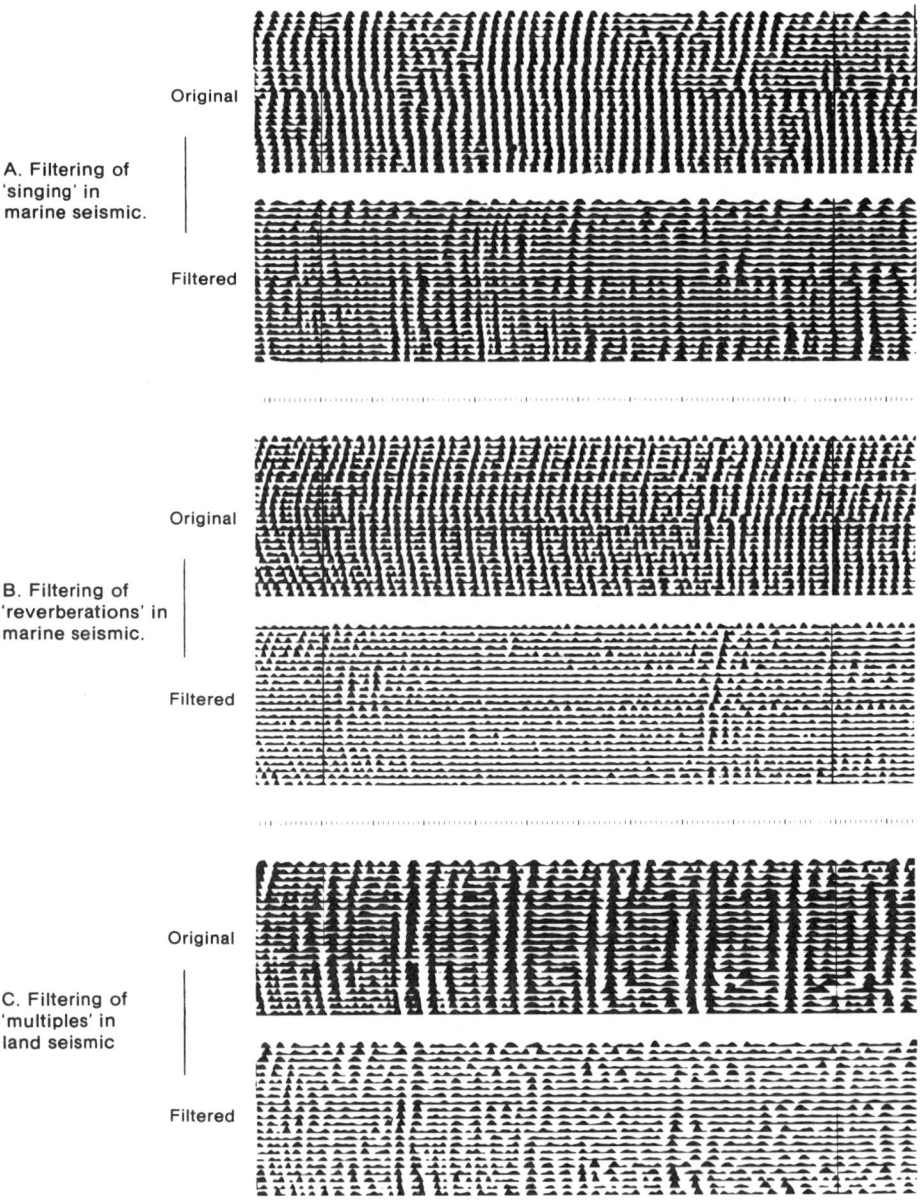

A. Filtering of 'singing' in marine seismic.

Original

Filtered

B. Filtering of 'reverberations' in marine seismic.

Original

Filtered

C. Filtering of 'multiples' in land seismic

Original

Filtered

Figure 27a. Samples of filtering.

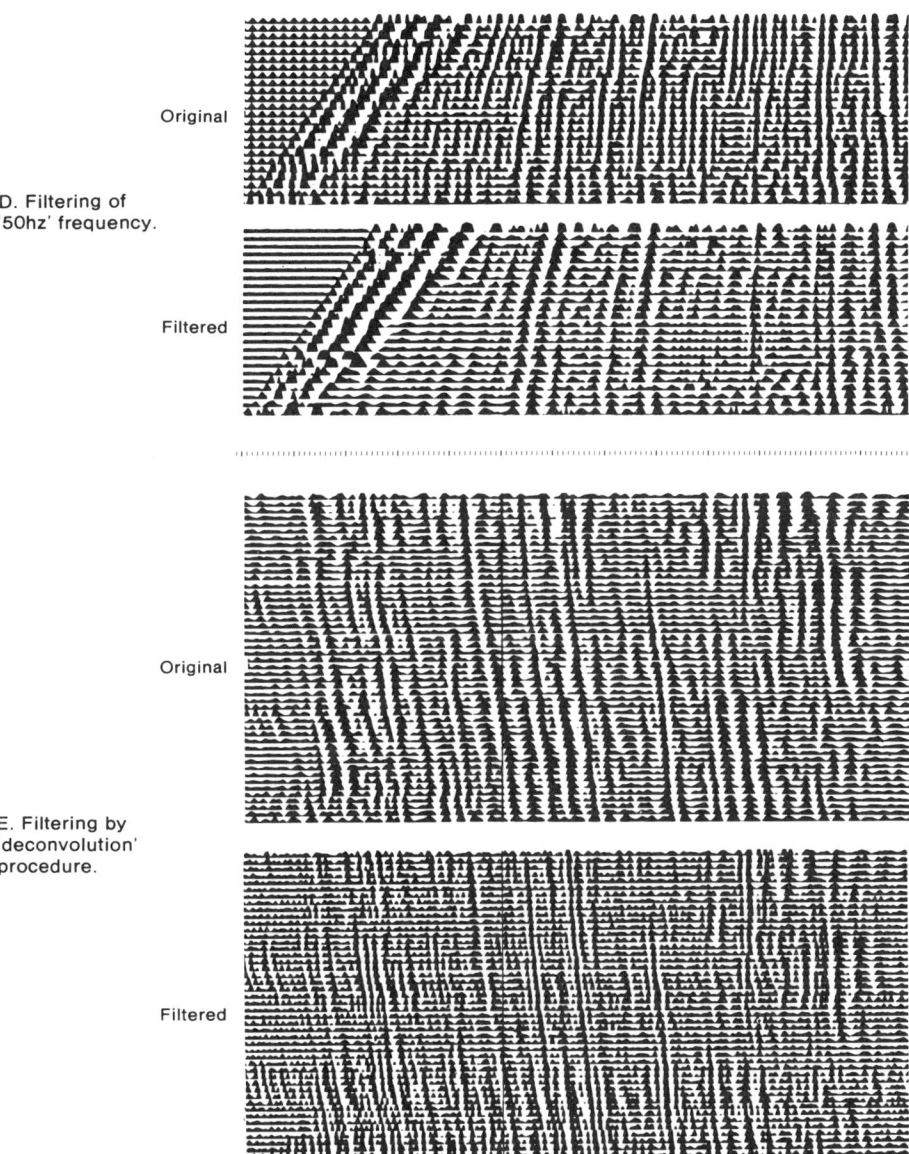

D. Filtering of '50hz' frequency.

Original

Filtered

E. Filtering by 'deconvolution' procedure.

Original

Filtered

Figure 27b. Samples of filtering.

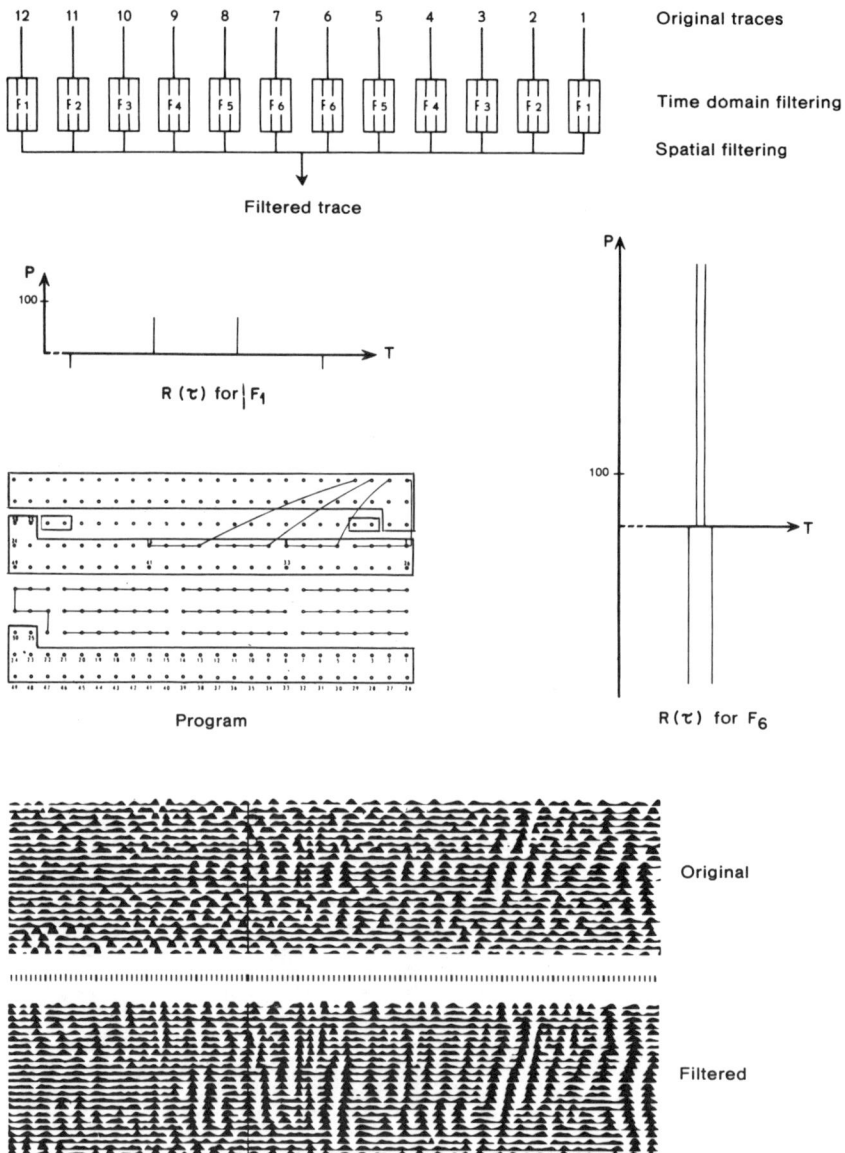

Figure 28. Filtering program board and results (fan filtering).

Use of two machines

The delay-line filter facilities which have just been described refer to the use of a single machine with 25 or 50 rings. These facilities are increased by dividing the equipment into two independent delay lines that can be used either in-parallel or in series.

In-parallel — Arranged in-parallel, the delay lines facilitate, among other things, the elimination of ghost arrivals, starting with two shots taken at different depths according to the method reported by Schneider et al (1964). Each of the two equivalent traces pass before summation into one or the other of the two delay lines containing the necessary operators.

In series — Taken in series, the delay lines allow us to conduct two successive filterings: either in time (filtering two different expressions of multiples, decon-volution and elimination of multiples, etc.) or spatially or in space/time and time (fan filtering followed by a deconvolution for example).

Implementation of the attenuation curve as a function of frequency

We have seen that implementation of time-domain filtering by the delay-line filter consists of convolving the seismic trace with the characteristic impulse re-sponse of the filter introduced in sampled form. We are otherwise accustomed to judging the effect of a filter by examining its action for each value of frequency, in other words to characterize it by cut-off values and slopes on a curve evaluating the attenuation as a function of frequency. This curve is easily obtained through the delay-line filter. Once the operator and program are in service, it is sufficient to replace the seismic trace at the input by a sinusoid coming from a generator capable of supplying 1 to 200 hz, and transmitting the output of the summation amplifier to a voltmeter. The filter response in the frequency band swept is ob-tained in this way.

This operation, which allows us to have the attenuation curve as a func-tion of frequency by a means other than calculation, constitutes a working test of the equipment.

Chapter 5

OPTICAL CORRELATION*

A. FONTANEL

An optical correlator was devised and perfected in 1962 at the French Petroleum Institute. Its use by geophysicists is so firmly established and it performs such important daily service that it would actually be difficult to consider the processing of seismic information, particularly marine work, without the valuable contribution of correlation studies. We are going to describe this device and its performance and then proceed to review its main applications.

Linear filters have been used very extensively in geophysics for several years. Autocorrelations, crosscorrelations, convolutions, and deconvolutions[4] are daily operations in seismic data processing. Therefore the IFP uses a simple device which performs the first two operations in an inexpensive and rapid manner. This device is the optical correlator for seismic analysis (CTS 1).

We shall consider the application of correlations only within the framework of reflection seismology, which is by far the most used geophysical method throughout the world. Beyond the previous considerations, there is another reason which makes correlation indispensable to the geophysicist. These are the problems posed by marine seismic work. The search for petroleum is turning more and more towards exploration of the continental shelf, and in this environment geophysics has a chosen place; surface geologic studies here are almost impossible, at best very fragmentary. By contrast, seismic methods are readily applicable here. A land crew produces only 3 or 4 km of profile per day, while at sea 60 or 80 km are realized during the same time.

But if, as we have just said, the implementation of shots and profiles is relatively easy, by contrast the water layer which covers the sedimentary section often enters into resonance through the shot effect and produces extremely disturbing reverberations. In this case, autocorrelation of the recordings is indispensable in determining the dereverberation filter operators.

THE SEISMIC OPTICAL CORRELATOR

Principle

To perform an optical correlation, we can choose to operate either in the frequency domain by taking the product of the spectra of the two functions directly or in the time domain; but in either case it is necessary to produce an appropriate photographic representation of the functions to be correlated.

---

*Paper presented at the meeting of the "Centre d'Etudes Théoriques de la Détection et des Communications," Paris, December 1964.

[4] The correlator was used for the determination of the operators.

Correlations in the frequency domain with coherent light[5]

We now employ the properties of diffraction at infinity, and as will be seen, the spectral product of two functions will be obtained directly. We can then return to the time functions still by optical means using an inverse Fourier transform.

Correlations in the time domain with coherent light

In this case we operate on the time functions themselves. With the first method the resulting correlation is displayed as differences in illumination in the focal plane, illuminations which express the values of the correlation function sought at each point (x, y), of the plane.

In contrast, the results of the second method are displayed as variations in electric current at the output of a photoelectric cell, which are simple to record, either on paper or on magnetic tape. It is mainly this correlation process that we will discuss here.

Consider two functions $s(t)$ and $y(t)$ to be correlated in the window $T = t_2 - t_1$ and let $g(t)$ be their autocorrelation function:

$$g(t) = \int_T s(t) \cdot y(t + \tau) \, dt.$$

The mathematical expression itself shows that there are three operations to perform:

Term by term multiplication of the two functions for a given shift τ — For this operation, the functions are transcribed on movie film by an appropriate display. The display should be such that the fraction of light flow which crosses both films superimposed in a uniform beam be linearly related at each portion of the films to the product of the two functions at the corresponding point. We shall see that several types of possible display exist.

Continuous shift of two functions — For this, one of the two films is fixed, the second is displaced over the first. This shift is performed in a continuous manner by transport drive with a broad range of speeds.

Integration of partial products over the whole interval under study — The light beam which has traversed the two film ensemble is made to converge on a photoelectric cell. After amplification, the current from the cell may be recorded either on magnetic tape or photographic paper by means of a recording pen.

The diagram of the optical correlator principle is very simply inferred from previous considerations (Figure 29).

---

[5] This topic will be treated briefly here.

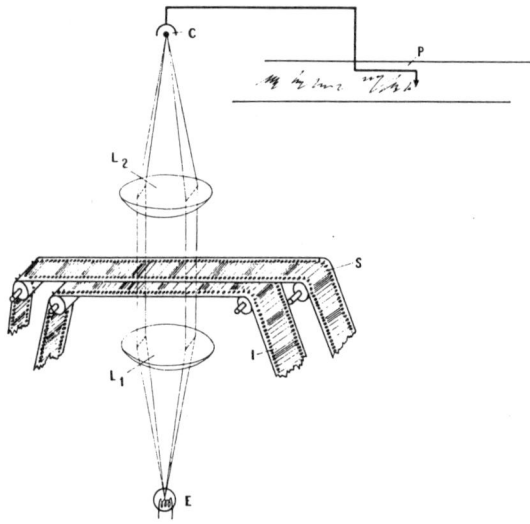

Figure 29.   Schematic diagram of optical correlator.

Description

The equipment comprises: a luminous source of small dimension (3 mm) and constant intensity (E), two large-diameter (30 cm) lenses $(L_1$ and $L_2)$ located on either side of the films and which form an image of the source on the photoelectric cell whose amplified current may be recorded on an ordinary commercial camera (Figures 29-31). For our purpose, we use either an x-y plotter or a Polaroid-type camera that registers the spot of a deflecting mirror galvanometer (Figure 31). An output is also provided to facilitate magnetic tape recordings.

In these different systems the progress of recording is always subject to the monitoring of the moving film.

The source is at the focal point of the first lens, the light beam crossing the films is therefore collimated. In addition, a rectangular stop between the two lenses intercepts light rays crossing portions of the films which are of no interest in the operation.

Performance and characteristics

We proceed initially to the transfer of the signal to be processed on 35-mm or 70-mm movie film. For 35-mm film this operation is carried out with a special camera (CT 1) which automatically records in sequence signals emanating from different magnetic tape tracks. The spools can store 120 m of film. This camera needs only very simple optics (in view of the so-called "wide-trace" display). It allows for three reel speeds: 38 cm/sec, 19 cm/sec, and 9.5 cm/sec.

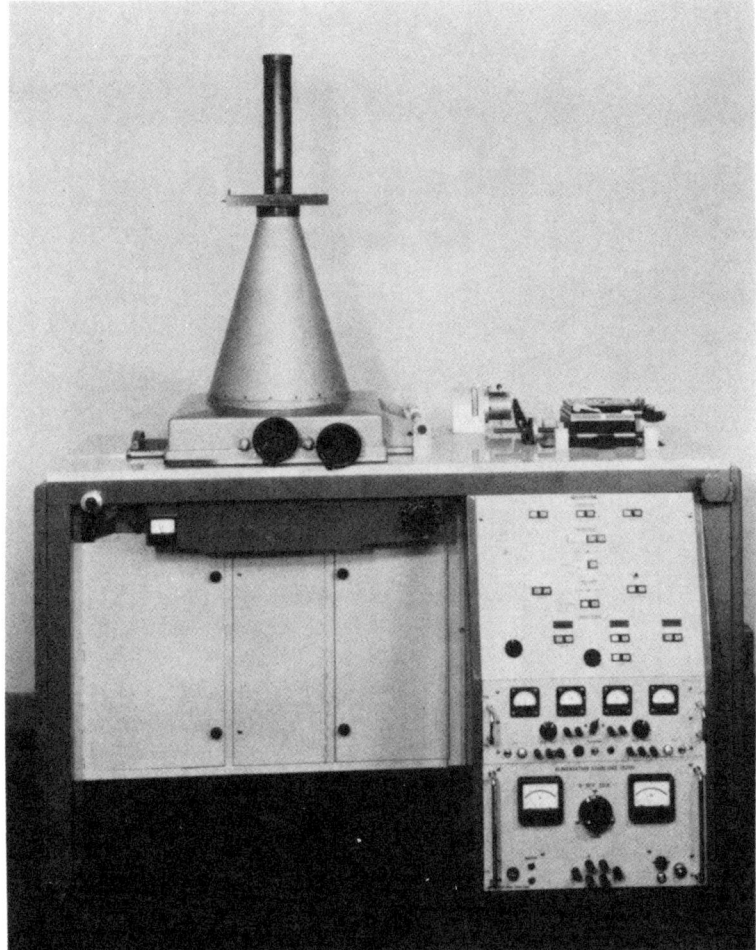

Figure 30.   Photograph of optical correlator.

As for the 70-mm film, no special camera has been constructed, but a typical seismic camera may be used.

The shift speed of the moving film in the correlator is governed by two factors:

— the passband of the recorder.[6]

— the desirable scale of the correlation display for a given recorder.

By means of two motors we can obtain the following speeds of the moving film:

a) 35-mm film: slow speeds: 1 mm/sec, 2 mm/sec;

---

[6] The normal seismic frequencies range from 5 hz to 150 hz. The passband of the correlator itself goes from 0 to 800 hz.

Figure 31.   Close-up of optical correlator.

35-mm film: high speeds: 23.7 mm/sec, 47.5 mm/sec, 95 mm/sec;

b) 70-mm film: slow speeds: 2 mm/sec, 4 mm/sec;

70-mm film: high speeds: 47.5 mm/sec, 95 mm/sec, 190 mm/sec.

A reversing device facilitates shifting the film in either direction; this is very practical in the case of 70-mm film. In fact, if for example several traces have been displayed side by side, the diaphragms permit trace selection and by successive advancing and reversing, the auto- or crosscorrelation series can be performed without having to displace the reference film.

Recording speeds of the correlation function also depend on the recorder employed.

The tracing table (spot following system) accomodates continuous variations of recording speed from 1.5 mm/sec to 12.5 mm/sec. Its passband is limited
to 1.5 hz upwards; it is therefore employed with slow correlator speeds.

The flat Polaroid film galvanometer recording system permits use of high
correlator speeds since the galvanometer passband is 0-175 hz.

By means of a gear box, 24 ratios of seismic film speed to that of Polaroid
recording film are possible, which permit a minimum correlation display of 150
msec, or a maximum of 15 sec with different intermediary gear.

Dynamic range

The dynamic range permitted by the correlation system — that is, including
the trace playback camera, the film display and the correlator itself — is 54 db. It
has been measured by performing:

a) on the one hand, the autocorrelation of two films through the complete
process of development and on which no signal has been introduced, in order to
obtain the noise threshold;

b) on the other hand, the autocorrelation of a sinusoid, which represents
the maximum signal obtainable in the course of an autocorrelation.

In Figure 32, a correlation obtained by the correlator may be compared with
that obtained by means of a digital computer after trace sampling. The window
correlated is of one-second duration and the operation performed is a pseudo-
correlation; in effect, a one-second window of the trace under study has been
chosen and has been correlated with the whole trace; this explains why the curves
obtained are not perfectly symmetrical.[7]

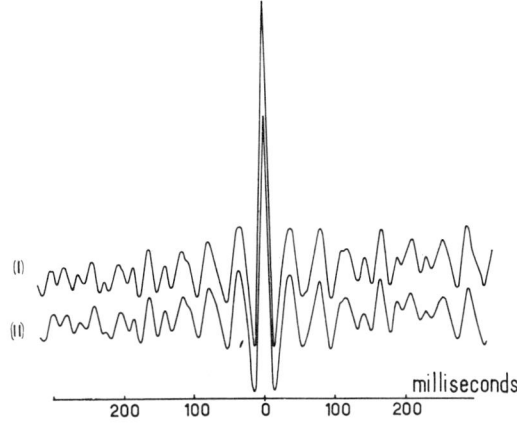

Figure 32. Convolution of a one-second sample with the whole original
trace (pseudo autocorrelation). (I) Computed; (II) optical correlator.

---

[7] In all examples which will be presented in the text, only puseudo-correlations of seismic traces
will be considered.

Principle of multiplication and optical integration

Multiplication (general case) — Consider an element of photographic film located in the P plane with coordinates x, y. It possesses a transmission coefficient $T_1 = T_1(x, y)$.

If this film receives a uniform light flow of intensity I_0, it transmits I:

$$I(x, y) = I_0\, T_1(x, y).$$

If a second film $T_2(x,y)$ is superposed, the distribution of light intensity in the emergent beam is given by:

$$I(x, y) = I_0\, T_1(x, y) \cdot T_2(x, y).$$

The product of the two functions represented by $T_1(x,y)$ and $T_2(x,y)$ has therefore been realized at each point.

Integration — In order to integrate along the entire length L of the films, it is sufficient to focus the emergent beam on a photoelectric cell. We then have:

$$I(x, y) = I_0 \int_L T_1(x, y). \ T_2(x, y)\ dx, dy.$$

The general case of two-dimensional functions has been considered here.

Different photographic presentations of seismic traces

Several types of photographic displays can be used to effect, in a more or less approximate manner, multiplication and consequently optical correlations.

One presentation, however, variable density, allows us to obtain an exact correlation; unfortunately it is difficult to use in practice. That is why we resort to the so-called "wide trace" presentation, which is very easy to produce and which enables us to obtain a correlation close to the exact one.

Variable density — Variable density consists of representing the different amplitudes of the function by the variations in transmission coefficient of the film. We are therefore conforming in this case exactly to the conditions of the optical multiplication principle previously described.

But in actual fact it is extremely difficult to effect a linear relation between the transmission coefficient and the function values. Indeed, the relation

$$T = k\, L^{-\gamma}$$

exists between transmission coefficient T and the lumination L.

L, the lumination, varies as the representative function, γ is the slope of the emulsion characteristic, and k is a coefficient of proportionality. It is therefore necessary, starting with the negative film to obtain a positive film with a transmission coefficient T'

$$T' = k' \, L'^{-\gamma'},$$

γ' being the slope of the emulsion characteristic.

It is also required that in order for the variable density display to respond to the conditions stated above.

$$\gamma\gamma' = 1.$$

This condition is very difficult to achieve, at least on large lengths of film.

In any case, even if this photographic requirement were met, the variable density presentation adds a constant, since negative light does not exist. In fact, the transmission coefficients are of the form

$$T_1(t) = T_0 + af(t),$$
$$T_2(t) = T_0 + ag(t).$$

where T_0 is the transmission coefficient of medium grey, that is the zero value of the function (seismic trace at rest); and a is the amplitude function assigned to the trace at the time of reply; and we can always consider it equal to 1. We then have

$$T_1 = T_0 + f(t),$$
$$T_2 = T_0 + g(t).$$

If I_0 is the light flow furnished by the source, the amount of light crossing a portion of the films, for a given shift τ, is:

$$I_0 \cdot \left[T_0 + f(t)\right] \cdot \left[T_0 + g(t+\tau)\right]$$
$$= I_0 \left[T_0^2 + T_0 f(t) + T_0 g(t+\tau) + f(t) g(t+\tau)\right].$$

The first two terms are constant during correlation and can easily be removed by electrical filtering during recording.

As to the third term, it varies with frequencies close to those of the correlation; furthermore, its amplitudes, weighted by the coefficient T_0, are generally at least equal to those of the correlation (and sometimes greater). It is therefore absolutely necessary to eliminate it. This is possible by means of a certain trace arrangement on the films but it is an additional difficulty due to the variable density. We can, for example, divide each film into two equal tracks and display the normal trace on the upper track and the same trace with reversed sign on the lower track:

    1st film   (upper track: $T_1 = T_0 + f(t)$ 2nd film $T_2 = T_0 + g(t)$

    1st film   (lower track: $T_1' = T_0 - f(t)$ 2nd film $T_2' = T_0 - g(t)$.

    First-degree terms are eliminated in correlation and there remains:

$$\varphi(\tau) = 2\,T_0^2 + 2\int_{t_1}^{t_2} f(t).g(t + \tau)\,dt,$$

that is, the exact correlation plus a constant.

For the different reasons just studied, the difficulty in correct photographic presentation, and the very great importance of certain error terms, we have been led to employ a novel display termed the "wide-trace" display (Continental Oil Co., 1960).

"Wide-trace" display — The different amplitudes of the function are represented here by oscillations along the main axis of an opaque film strip of constant width (Figure 33).

Figure 33.   Wide-trace representation.

It is well known that in this case, the integral M of the absolute value of the difference (or sum) of the two representative functions $f(t)$ and $g(t)$ is obtained with an optical correlation:

$$(M) = \int_{t_1}^{t_2} |f(t) \pm g(t + \tau)|\,dt.$$

It will be seen that given the seismic trace properties, a ratio close to 1 generally exists between the exact correlation function (P) and the previous function (M).

Let us indicate from now on that this presentation, in contrast to the variable density, is easy to obtain since there is only black and white to display on the film.

We have seen that in the case of variable density an error term exists due to mean value fluctuations of the function being shifted, which is troublesome for it is weighted by the transmission coefficient of medium gray.

In the case of wide-trace presentation another error term is substituted with the form (see Appendix):

$$\int_{t_1}^{t_2} g^2(t + \tau)\,dt.$$

For seismic traces, it is of lower frequency than the correlation and can therefore be eliminated to a large extent by electrical filtering during recording.

Furthermore, for the autocorrelation of a function $f(t)$ over a window of length T, the curve obtained through the integral of the absolute value of the difference, namely:

$$\int_T |f(t) - f(t + \tau)| \, dt$$

presents a low frequency of sawtooth form with period $2T$. The correlation values are represented by the sawtooth fluctuations around the low frequency, the latter is removed by electrical filtering during recording. In contrast, when the entire shift window is involved in the case of a pseudoautocorrelation, the curve oscillates about a level that will be defined in the Appendix.

a. *Justification for employing the wide-trace display* — It is demonstrated (see Appendix) that, given the seismic trace properties such as zero mean, a ratio close to 1 generally exists between the exact correlation function (P) and the function representing the integral of the absolute value of the difference (M), the two functions being reduced to the same mean scale.

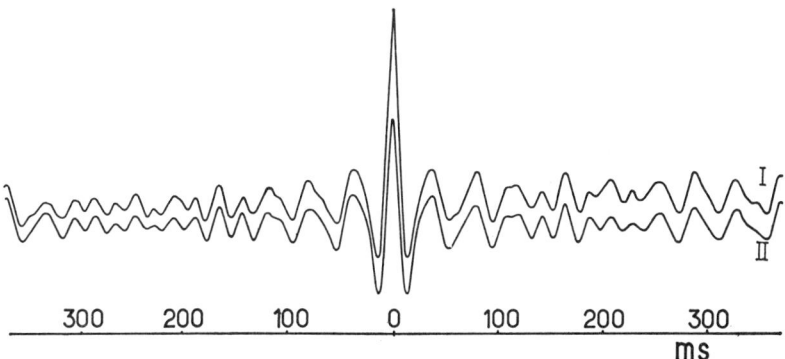

Figure 34. Comparison of the exact autocorrelation (I) (variable density) with the approximation (II) (wide-trace). Curves obtained, after trace sampling, by means of a digital computer.

That is, if the noncentral values of the correlation are well represented by the integral of the absolute value of the difference, the magnitude of the central peak is exaggerated and can even be double its real value. This is readily observed on Figure 34 where curve *I* relates to function M and curve *II* depicts function P.

Between the curves previously defined as (P) and (M), we have the relation:

$$E = \frac{(P)}{(M)}$$

$$= \frac{2}{1 + \sqrt{1 - \dfrac{P}{P_0}}}$$

where P is the autocorrelation amplitude at any local point and P_0 is the amplitude of the central peak (Figure 35). Actually, this relation is only approached and curve (M) often displays low-frequency oscillations due to the fact that the mean square of the shifting function, in the interval under study, is not constant.

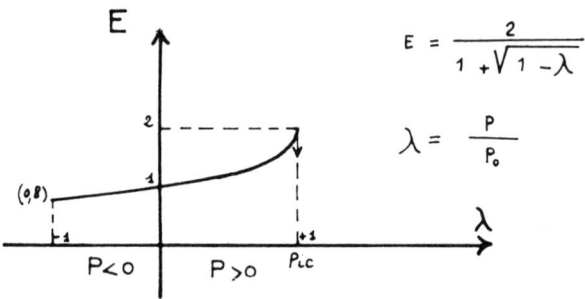

$$E = \frac{2}{1 + \sqrt{1 - \lambda}}$$

$$\lambda = \frac{P}{P_0}$$

Figure 35. Ratio of the exact correlation P to the approximation M as a function of the normalized correlation P.

$$P = \int_{t_1}^{t_2} f(t)\, g\,(t + \tau)\, dt,$$

$$M = \int_{t_1}^{t_2} |f(t) - g\,(t + \tau)|\, dt.$$

These low-frequency oscillations are, of course, filtered and generally cannot be seen on the correlation records.

Finally, the wide-trace display favors positive autocorrelation values and reduces negative amplitudes slightly; but of special interest is the fact that the central peak is doubled, which is advantageous in certain cases of seismic prospecting such as vibrator studies.

EXAMPLES OF USE

Ghosts in land seismic work

For the purpose of ghost investigation we have conducted systematic autocorrelation of traces (Figure 36) relating to shots located at different depths (Lindsey, 1960). Actually, part of the energy in a buried shot may be reflected at the air-ground interface before penetrating the geological section.

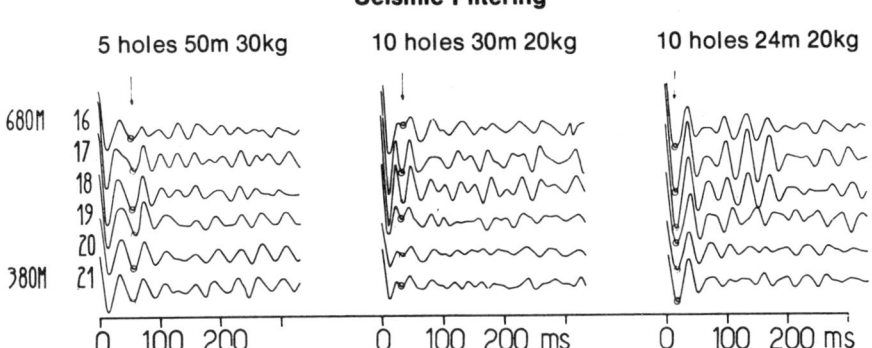

Figure 36.   Ghosts in land seismic, 60 m between traces.

This gives rise to a secondary impulse, termed a ghost, delayed with respect to the primary energy. The delay is of course a function of shot depth and the velocity of wave propagation.

The position of the ghost indicated by an arrow is seen in the diagram. On the left, the shot was 50 m deep, then 30 m and 20 m; its displacement can be followed. Sometimes it arrives in-phase with the amplitudes of the correlation (20-m and 50-m shot), sometimes it arrives in opposition to them (30-m shot).

We have here a means of displaying the ghosts which often interfere and particularly of measuring their delay with respect to the primary reflection. Filters can therefore be applied to remove them.

Inverse filtering of marine seismic data

Shallow water depth: ringing — It quite often happens that marine records are strongly disturbed by resonance due to the water layer to the extent of making the recording uninterpretable. It is important to know exactly this frequency of "ringing" for the purpose of accurately determining the filter to be used. It can be seen in Figure 37 that the autocorrelation brings it out very well; the former, in fact, resembles a damped sinusoid.

The ringing frequency is 40 hz here. It would now be necessary to know what the energy carried by this 40 hz is relative to adjacent frequencies for the purpose of establishing the depth of the filter; this will be the object of harmonic analysis, also possible with the optical correlator as will be seen later.

Reverberations (repetitions) — When the water depth becomes sufficiently great; the wave trains that are reflected several times within the water appear on the recording as distinct arrivals called reverberations. A pseudoautocorrelation of a trace displaying these reverberations may be seen in Figure 38.

On either side of the central event, arrivals of alternate sign are clearly distinguished which correspond to multiple reflections between the water bottom

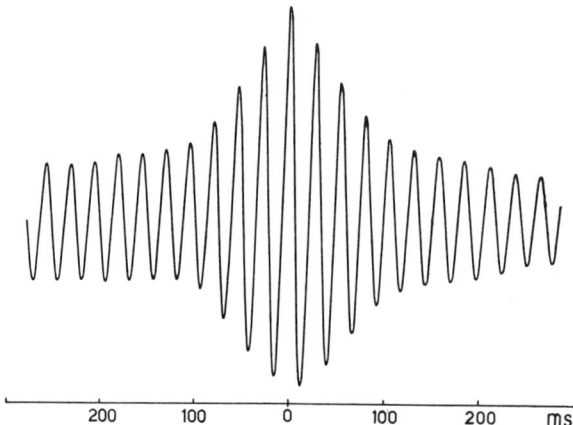

200   100   0   100   200   ms

Figure 37.   Pseudo-autocorrelation of a marine seismic trace, in shallow water.

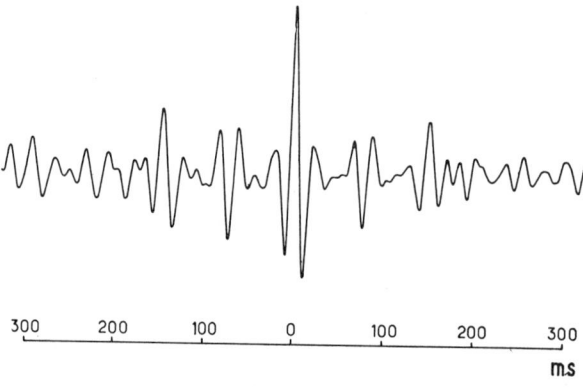

300   200   100   0   100   200   300

ms

Figure 38.   Pseudo autocorrelation of a marine seismic trace, in deep water.

and the surface; they make picking of horizons difficult. Knowledge of the autocorrelation and in particular of the delay time between these different energy arrivals and their respective sign is very useful for the determination of an effective filter.

For the purpose of designing a filter, assemblages of autocorrelations which permit following the development of the phenomenon are realized in this manner, to see at which instant it is necessary to modify the filter characteristics employed. It is thus noticed that quite often resonance is not caused by the bottom as indicated by the depth sounder. It is sometimes advantageous to take the sum of autocorrelations of several seismic traces. In this way a clearer view of the reverberation phenomenon can be obtained. This operation is very easily achieved by employing multitrace film; fluctuations recorded by the photoelectric cell correspond to the sum of the autocorrelations of individual traces.

Deconvolution

In order to deconvolve seismic traces, it is essential to know the seismic pulse. One of the main problems in deconvolution is the lack of information about this pulse. But the autocorrelation function provides a means of calculating an approximate wavelet and from this approximation to define an inverse deconvolution operator. Figure 39 accurately portrays a correlation made for the purpose of calculating a deconvolution operator (nonfiltered trace). It likewise illustrates the use that can be made of the autocorrelation to control filter choice.

Control of filter selection

Visual judgment of seismic records on the effectiveness of two seismic filters is something essentially qualitative. With correlation we have the means of specifying our choice. The upper curve in Figure 39 represents the pseudoautocorrelation of an unfiltered trace. In the center of the figure we show the trace autocorrelation after application of an electrical filter with 60–120 hz passband. At the bottom is displayed the pseudoautocorrelation of the trace filtered by a deconvolution operator. It can be seen how very good the correlation is between the different events on these two curves; however, the last curve shows much higher frequencies and greater resolution.

Static corrections determined by the correlation method

One method of determining static corrections is to note the shift on each trace required to align a reflection on a seismic record. It is often difficult and

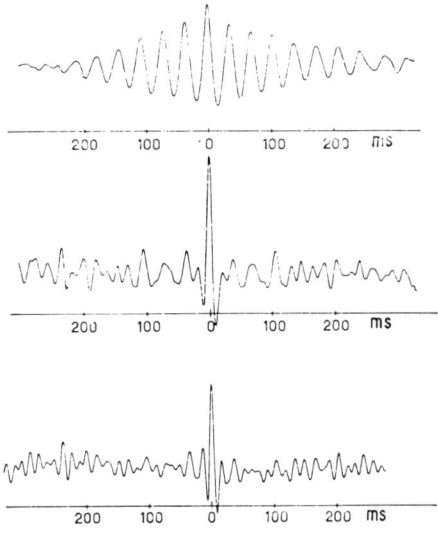

Figure 39. Pseudoautocorrelations of a nonfiltered trace (top), same trace filtered with a passband of 60 to 120 hz (center), and trace time-delay filtered (bottom).

not very accurate to determine arrival times on the seismic films for this purpose. A less subjective method consists of crosscorrelating different traces. If an amplitude stronger than adjacent ones exist in the crosscorrelation, the adjustment relative to the two traces should be made at the corresponding shift time; in this way the picking of the arrivals is determined sequentially in crosscorrelating traces to be aligned.

These different seismic traces are arranged side by side on two 70-mm films, so that when the two films are placed in contact with each other, trace 1 overlies trace 2, etc. The correlator diaphragms allow alternate selection of each of these trace pairs and by successive adjustments, the entire series of autocorrelations, which are recorded below each other with the necessary adjustment, is accomplished. A picking accuracy of one millisecond is easily attained.

Another solution also consists of crosscorrelating a certain number of traces, with the same reference trace; this can be seen in Figure 40 where traces 1 to 12 have been correlated with trace 10.

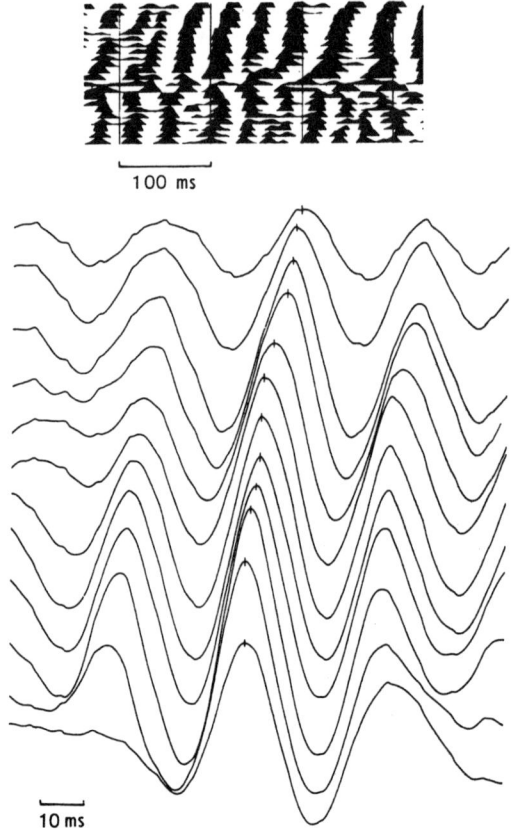

Figure 40.   Calculation of static corrections by correlation of traces 1 to 12 with trace 10. Original record at top.

Harmonic analysis by optical correlator

If a function is correlated with sinusoids of different frequencies, for each of these frequencies a sinusoid is obtained at the correlator output. The sinusoid amplitude represents the absolute value of the spectrum for the function at the frequency considered.

Operation

The seismic traces analyzed are generally of one-sec length. Correlation is accomplished in this case by a roll of film displaying a sinusoid whose frequency changes linearly with a variation slightly less than one hz/sec. The end frequencies are those of the seismic band: 5 hz and 100 hz. The harmonic analysis is then obtained in continuous fashion from 5 to 100 hz with an accuracy in the order of one hz.

The fast speeds of the correlator may be utilized; it is then necessary to operate from 30 to 60 sec, according to the frequency band studied, in order to carry out the harmonic analysis of a seismic trace. Figure 41 displays this type of spectrum obtained on Polaroid film. In Figure 42 a series of autocorrelations can be compared with the spectra, corresponding to the same seismic traces.

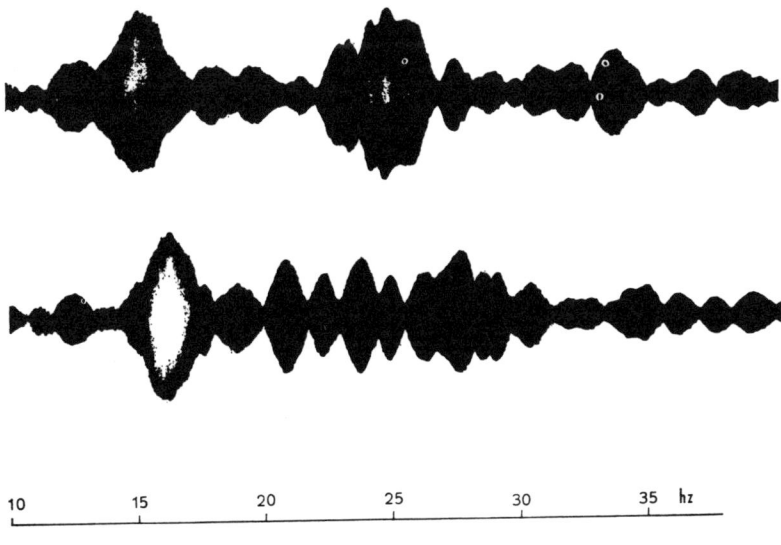

| 10 | 15 | 20 | 25 | 30 | 35 hz |

Figure 41.   Spectra of seismic traces, from the original correlator.

There is another field in which the geophysicist must resort to correlation; it is that of the vibroseis (Continental Oil Co., 1960) (see Chapter 8). This prospecting method actually consists of transmitting into the ground by means of a vibrator, a known long signal with flat spectrum, and whose autocorrelation is narrow. The type of signal selected is generally a chirped sinusoidal signal (e.g., 20 to 90 hz) of 6 to 8 sec duration.

Optical Correlation

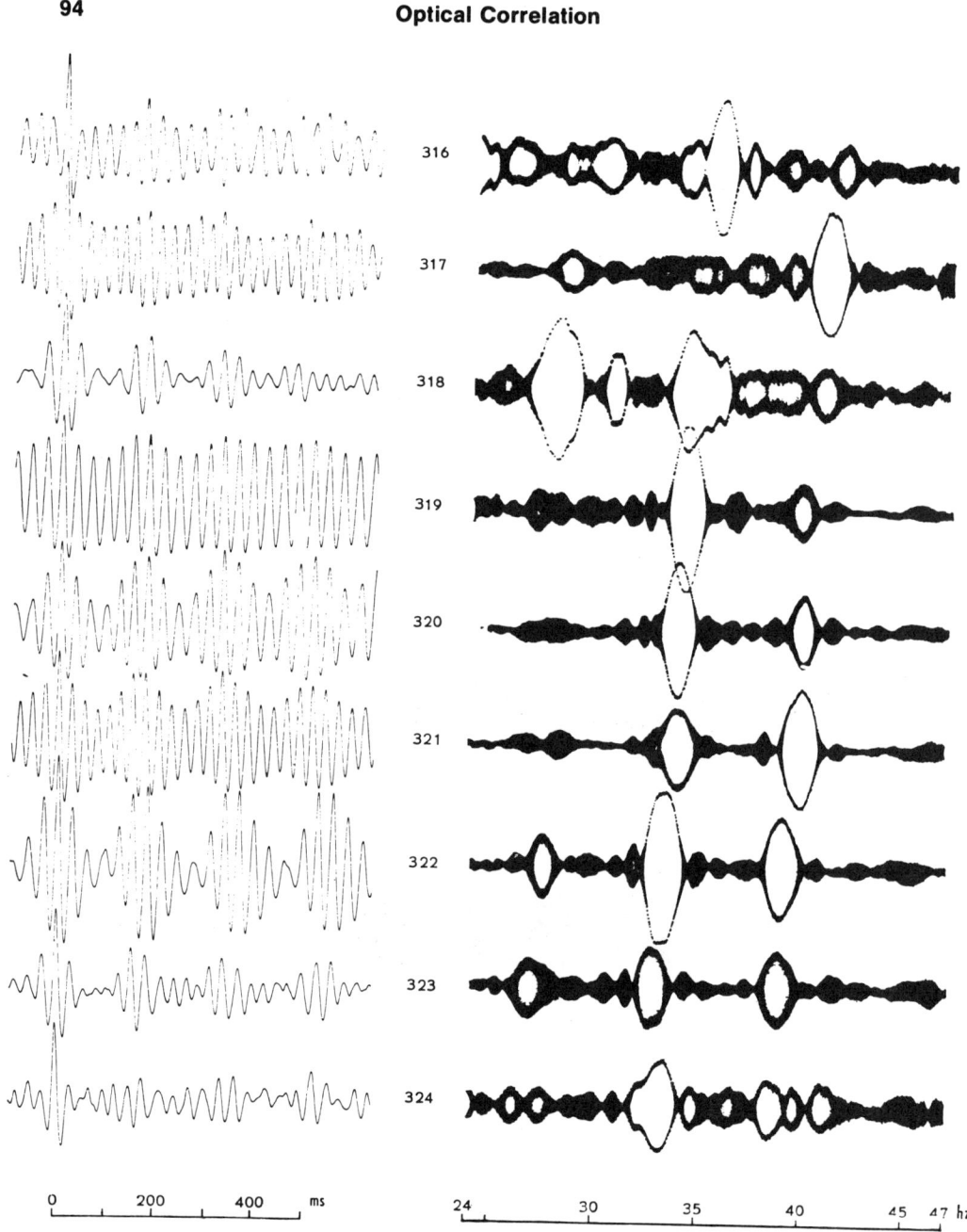

Figure 42. Comparison of autocorrelations (left) and spectra (right) of a series of trace 13's at a succession of shotpoints. Recording bandpass 18-75 hz.

Recording is done on the surface in the same way as for conventional seismic work, but it is necessary to correlate the recorded signal with the emitted signal in order to recover what would have been obtained if a sharp impulse had been transmitted to the earth.

The French Petroleum Institute has designed a special optical correlator for field use; correlation is performed in real time with this equipment.

PRINCIPLE OF OPTICAL CORRELATOR EMPLOYING THE PROPERTIES OF DIFFRACTION AT INFINITY

Theory

The diffraction pattern at infinity of coherent light by an object $f(x,y)$ represents the spectrum of the object, namely: $F(u,v)$. One can readily verify this at home by looking at a distant source of light (street lamp) through a fine-woven curtain. A series of blurs arranged in the form of a cross is seen, representing a portion of the two-dimensional Fourier transform of the curtain mesh.

A convergent lens is employed to bring back the diffraction pattern from infinity to a finite distance. The spectrum then forms in the conjugate plane of the source relative to the lens.

The object should be illuminated with coherent light, employing either a laser or a monochromatic point source.

In this case, to accomplish the convolution of two functions $f(x, y)$ and $g(x, y)$ it is sufficient to position a photographic plate displaying the spectrum of $g(x, y)$, that is, $G(u,v)$, in the spectral plane (image focal plane of the previous lens). The product of the spectra is achieved in this manner, that is,

$$\varphi(u,v) = F(u,v) \cdot G(u,v).$$

It is now necessary to return to the functions x, y. For this purpose the object image is formed by means of an optical system employed, according to the classical laws of geometrical optics. We then obtain:

$$\varphi(x,y) = f(x,y) * g(x,y).$$

Convolution of two functions f and g thus has been accomplished without having to shift one of the functions relative to the other (Vander Lugt, 1964).

Using this principle, a very simple filtering procedure consists of removing from the spectrum of the object function light which corresponds to arrivals of parasitic energy. It suffices to mask, as desired, the areas in the focal plane. We can perform frequency filtering or fan filtering in this fashion (Jackson, 1965; Dobrin et al, 1965).

One thus can effect, in a single operation, two-dimensional filtering (or convolution) of an entire record section of at least 500 traces. For example, fan filtering could be done in this fashion. This correlation procedure is therefore flexible and its execution simple. Figure 43 shows an example of two-dimensional spectrum of a record section.

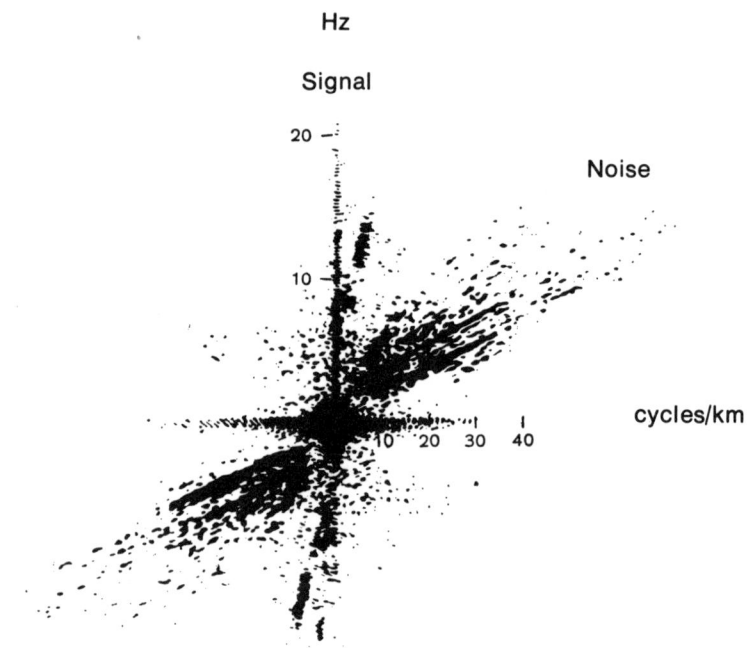

Figure 43.   Two-dimensional spectrum of a noise shot.

CONCLUSION

As we have just seen, two optical correlators of different types are now used (1964):

1. First, a correlator using incoherent light, intended for detailed analysis of each trace or a small number of traces simultaneously, in order to accurately determine the filter operators which will be applied elsewhere.

2. Second, a correlator based on diffraction properties at infinity of coherent light, which not only permits the display of the two- or one-dimensional spectrum of a seismic section, but also the filtering of this section and the capability of viewing on a television screen and photographing the results of filtering.

The correlators complement each other, for if the one permits a simultaneous overall view of the seismic section along the frequency and wavenumber axis, the detailed structure of the seismic trace is attained with the other by virtue of its analytical resolution. Now for problems of increasing complexity which present themselves to the geophysicist, linear filtering alone provides for solution at the present time. In this context the optical procedures are extremely interesting, for by their simplicity and flexibility they often permit producing elegant solutions at low cost.

APPENDIX

(Contributed by M. La Porte)

Relation between exact correlation (variable density)
and approximate correlation (wide trace)

The demonstration is made for the case of pseudoautocorrelation[8] which is most often performed in the optical correlation of seismic traces. We will call $f(t)$ the window of length T and $g(t)$ the whole of the same seismic trace.

Let

$$M = \int_{(T)} |f(t) - g(t + \tau)| \, dt \qquad \text{(wide trace display)},$$

$$P = \int_{(T)} f(t) \cdot g(t + \tau) \, dt \qquad \text{(variable density display)}.$$

It is known that if a is a gaussian variable of zero mean, there exists the relation:

$$\text{Mean} \, |a| = \sqrt{2/\pi} \ \sqrt{\text{Mean} \, a^2}.$$

We shall assume that the seismic traces possess these properties; we can then write

$$\text{Mean} \, |f - g| = \sqrt{2/\pi} \ \sqrt{\text{Mean} \, (f - g)^2}$$

or in another form:

$$1/T \int_{(T)} |f - g| \, dt = \sqrt{2/\pi} \ \sqrt{1/T \int_{(T)} (f - g)^2 \, dt}. \tag{1}$$

Setting:

$$\int_{(T)} f^2 \, dt + \int_{(T)} g^2 \, dt = A^2 \,,$$

equation (1) becomes:

$$(M) = \int_{(T)} |f - g| \, dt = A \sqrt{\frac{2T}{\pi}} \ \sqrt{1 - \frac{2}{A^2} \int_{(T)} fg \, dt}. \tag{2}$$

The term $2 \int_{(T)} fg \, dt$ is generally small in relation to the term $\int_{(T)} f^2 \, dt + \int_{(T)} g^2 \, dt$

---

[8] A pseudoautocorrelation is the correlation of a certain seismic trace window with the whole of the same trace.

and we can write:

$$(M) \simeq A \sqrt{\frac{2T}{\pi}} \left(1 - \frac{1}{A^2} \int\limits_{(T)} f g \, dt\right)$$

$$(M) \simeq A \sqrt{T} \sqrt{2/\pi} - (\sqrt{T}/A) \sqrt{2/\pi} \ (P)$$

$$\boxed{(M) \simeq C_0 - k \ (P) \, .}$$

The negative sign appears, since the variations in M are opposite to those of P.

C_0 will be called the curve (M) constant level; it represents the continuous component that exists in the pseudoautocorrelation when the entire moving trace is engaged. At the instant of coincidence between $f(t)$ and $g(t)$, i.e., at the auto-correlation peak, the two segments $f(t)$ (fixed) and $g(t)$ (moving) are identical; their difference is therefore zero; at this point the curve (M) falls back to zero value[9]. The height of the C_0 level then represents the amplitude of the autocorrelation peak in the case of curve (M).

Comparison of autocorrelation peak: curves (M) *and* (P)

Let Γ_0 be the peak of (M) reduced to the scale of (P), both curves being re-duced to the same mean scale:

$$\Gamma_0 = \frac{C_0}{k} = A^2.$$

At the autocorrelation peak, we have

$$(P)_{\text{Peak}} = \int\limits_{(T)} f^2 \, dt \quad \text{and} \quad [(f\,(t) \text{ and } g\,(t) \text{ identical}].$$

Then

$$(P)_{\text{Peak}} = A^2/2 = \Gamma_0/2$$

$$(P)_{\text{Peak}} = M_{(\text{Peak})}/2.$$

With both curves reduced to the same mean scale, the peak given by the wide trace is hence twice as large as that of the variable density.

General case

Let us now examine how the two curves (M) and (P) compare at any point. For this, we are going to compare (P) with (M) reduced to its level C_0 and brought back to the same mean scale; let $\mathfrak{M}$ be this new curve, changed in sign:

$$\mathfrak{M} = - \frac{M - C_0}{k}.$$

[9] The two expressions M and P defined above are compared here, it being un-derstood that, in the case of optical correlation, a constant is added to these two functions since negative light does not exist.

Setting

$$E = \frac{\mathfrak{M}}{P} \cdot$$

We have, according to equation (2):

$$E = \frac{C_0}{kP} \left[1 - \sqrt{1 - \frac{2P}{A^2}} \right]$$

$$= 2 \frac{P_{\text{Peak}}}{P} \left[1 - \sqrt{1 - \frac{P}{P_{\text{Peak}}}} \right] \cdot$$

which gives

$$E = \frac{2}{1 + \sqrt{1 - \frac{P}{P_{\text{Peak}}}}} \cdot$$

This curve is drawn in Figure 35. When P is small in relation to P_{Peak}, we have $E \simeq 1$. This indicates that when the amplitude variations of the autocorrelation function are small with respect to the amplitude of the central peak, curve $(\mathfrak{M})$ and curve (P) are very similar.

Fluctuations in the level of (M)

The term C_0 is not absolutely constant as a function of the shift in $g(t)$:

$$C_0 = A \sqrt{T} \sqrt{\frac{2}{\pi}},$$

with: $A^2 = \int_{(T)} f^2 \, dt + \int_{(T)} g^2 \, (t + \tau) \, dt$; the last term is the one that varies.

$$C_0^2 = \frac{2T}{\pi} \left[\int_{(T)} f^2 \, dt + \int_{(T)} g^2 \, (t + \tau) \, dt \right]$$

$$2 C_0 \Delta C_0 = \frac{2T}{\pi} \Delta \left(\int_{(T)} g^2 \, (t + \tau) \, dt \right)$$

$$\Delta C_0 = \frac{T}{\pi C_0} \Delta \left(\int_{(T)} g^2 \, (t + \tau) \, dt \right)$$

$$= \frac{1}{A} \sqrt{\frac{T}{2\pi}} \Delta \left(\int_{(T)} g^2 \, (t + \tau) \, dt \right).$$

This term has lower frequencies than those of the correlation and is removed in large part by electrical filtering during recording of the correlation.

Chapter 6

INVERSE FILTERING IN THE CASE OF NORMAL INCIDENCE (PLANE WAVES)

CH. HEMON

Recordings made in land shooting, in the course of petroleum exploration by the seismic method, display the existence of vibration phenomena which takes place in three-dimensional space. The very manner of recording and display, made in the form of seismic sections, express these phenomena as though they had taken place in the vertical plane containing the line connecting shotpoint and seismometers, thus restoring the diagram to two dimensions. Even with this simplification, the theoretical description of the phenomena remains extremely complex and has not yet received a simple general solutuon; up to now this inverse filtering has been obtained only in the case of a single-dimension display.

After the outline of this solution, it will be seen how it facilitates presentation of linear and nonlinear filter forms whose application effectively enhance useful information and tend to discard progressively all that is not essential to the geophysical interpretation. Several methods will be described here that have been found useful without aiming to establish a complete catalog on what has been published on the subject. Likewise, the bibliography is not intended to be complete, but should provide a general insight into the problems posed by the inverse filter.

THEORY OF IMPULSE AND SYNTHETIC SEISMOGRAMS

The seismic trace recorded by the geophone represented as a function of time will be called impulse seismogram (IS) if it corresponds to a theoretical explosion giving rise to a Dirac impulse at the instant t_0 of the explosion.

This impulse $\delta(t - t_0)$ is zero for $t \neq t_0$ and

$$\int_{-\infty}^{+\infty} \delta(t - t_0)\, dt = 1.$$

The absolute value of its spectrum is equal to 1 whatever the frequencies may be. If, on the other hand, the explosion generates an impulse of very long duration with respect to the time sampling interval employed for the calculation, the recorded curve will be called synthetic seismogram (SS).

Computations are made with the following assumptions:
— the waves propagate at normal incidence;
— planes separating beds, within each of which velocity and density are constant, horizontal, and parallel;
— the wavefronts are plane and parallel to the surfaces of equal velocity and equal density.

— materials which constitute the subsurface are perfectly elastic and obey Hooke's law;

— absorption in the subsurface is neglected.

Use of numerical computation requires first that the curves of velocity and density be sampled, in order to replace the given heterogeneous medium whose properties may vary continuously, by a heterogeneous medium composed of a sequence of homogeneous beds. The fictitious homogeneous beds are such that each is traversed by the wave in the same time τ. The waves do not undergo change in any of the beds other than a delay, but there is a reflection and transmission at the contact between the beds.

The principle of the matrix product method comprises considering the change that passage through a bed and a contact imposes on the waves which vibrate the bed sequence. The equation of plane-wave propagation can be applied in each of the beds, given the assumptions made above.

If the wave amplitude is designated by u, the bed density by ρ, the velocity of wave propagation in this bed by V, this equation is written

$$\frac{\partial^2 u}{\partial t^2} = V^2 \frac{\partial^2 u}{\partial x^2},$$

where t is the time variable and x the distance traveled by the wave. It is known that if we set $u = u_0(x)e^{j\omega t}$, that is the waves are assumed to be stationary, the previous equation with partial derivatives becomes a differential equation:

$$\frac{d^2 u_0}{dx^2} + \frac{\omega^2}{V^2} u_0 = 0,$$

whose solution is

$$u = a\, e^{j\omega(t-x/V)} + b\, e^{j\omega(t+x/V)}.$$

This equation applies only for the interior of the given bed, the origin of x is to be taken as the top of the bed; the maximum value of x is equal to $V\tau$, V being the velocity of wave propagation in the bed.

Consider the interface m (Figure 44) which separates beds of order $[m-1]$ and $[m]$, characterized by densities and velocities $\rho_{m-1}, \rho_m, V_{m-1}, V_m$.

In each bed, the previous equation is written

$$u_{m-1} = a_{m-1}\, e^{j\omega\,(t-x/V_{m-1})} + b_{m-1}\, e^{j\omega\,(t+x/V_{m-1})}$$

and

$$u_m = a_m\, e^{j\omega\,(t-x/V_m)} + b_m\, e^{j\omega\,(t+x/V_m)}.$$

In these formulas a_{m-1} and a_m are the downgoing wave amplitudes of frequency $f = \omega/2\pi$, b_{m-1} and b_m are the returning wave amplitudes of frequency $f = \omega/2\pi$. At each discontinuity between homogeneous beds there should be con-

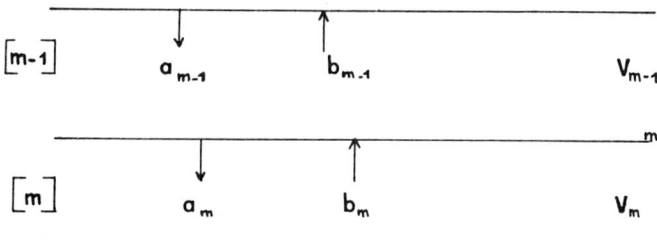

Figure 44.   Schematic diagram showing waves in the $(m-1)^{\text{th}}$ and m^{th} horizontal beds.

tinuity of particle movement and of normal stress whatever the time may be. At interface m, the first of these conditions is written as

$$[u_{m-1}]_{x=\tau V_{m-1}} = [u_m]_{x=0}$$

and the second

$$\rho_{m-1} V_{m-1}^2 [du_{m-1}/dx]_{x=\tau V_{m-1}} = \rho_m V_m^2 [du_m/dx]_{x=0}.$$

In explaining these two equations, we obtain, after dividing both sides by $e^{j\omega t}$:

$$a_{m-1} e^{-j\omega\tau} + b_{m-1} e^{j\omega\tau} = a_m + b_m$$

$$\rho_{m-1} V_{m-1} [a_{m-1} e^{-j\omega\tau} - b_{m-1} e^{j\omega\tau}] = \rho_m V_m [a_m - b_m].$$

This system of two linear equations in $a_{m-1}, b_{m-1}, a_m, b_m$ is conveniently written in matrix form:

$$\begin{pmatrix} a_{m-1} \\ b_{m-1} \end{pmatrix} = \begin{pmatrix} e^{-j\omega\tau} & e^{j\omega\tau} \\ e^{-j\omega\tau} & -e^{j\omega\tau} \end{pmatrix}^{-1} \begin{pmatrix} 1 & 1 \\ \dfrac{\rho_m V_m}{\rho_{m-1} V_{m-1}} & -\dfrac{\rho_m V_m}{\rho_{m-1} V_{m-1}} \end{pmatrix} \begin{pmatrix} a_m \\ b_m \end{pmatrix}.$$

By effecting changes of variables

$$z = e^{-2j\omega\tau} \text{ and } r_m = \frac{\rho_{m-1} V_{m-1} - \rho_m V_m}{\rho_{m-1} V_{m-1} + \rho_m V_m},$$

we obtain the relation:

$$\begin{pmatrix} a_{m-1} \\ b_{m-1} \end{pmatrix} = \frac{z^{-1/2}}{1 + r_m} \begin{pmatrix} 1 & r_m \\ r_m z & z \end{pmatrix} \begin{pmatrix} a_m \\ b_m \end{pmatrix}.$$

This equation is basic for it allows us to express the a and b of a bed as a function of those of another bed without these two beds being necessarily adjacent. In particular, a matrix equation can be written which relates the amplitudes a_1 and b_1 of the waves in the first bed and the amplitude a_N of the downgoing wave in the last bed. Assuming the former to be semi-infinite, there is no upward wave which implies that $b_N = 0$.

We have

$$\begin{pmatrix} a_1 \\ b_1 \end{pmatrix} = z^{-(N-1)/2} \prod_{m=2}^{N} \left[\frac{1}{1+r_m} \begin{pmatrix} 1 & r_m \\ r_m z & z \end{pmatrix} \right] \begin{pmatrix} a_N \\ 0 \end{pmatrix},$$

which can be written in the form:

$$\begin{pmatrix} a_1 \\ b_1 \end{pmatrix} = a_N \, z^{-(N-1)/2} \prod_{m=2}^{N} \frac{1}{1+r_m} \begin{pmatrix} P \\ Q \end{pmatrix},$$

P and Q being polynomials in z defined by the relation:

$$\begin{pmatrix} P \\ Q \end{pmatrix} = \prod_{m=2}^{N} \begin{pmatrix} 1 & r_m \\ r_m z & z \end{pmatrix} \begin{pmatrix} 1 \\ 0 \end{pmatrix}.$$

Setting

$$\lambda = z^{-(N-1)/2} \prod_{m=2}^{N} \frac{1}{1+r_m},$$

we have

$$a_1 = \lambda P \, a_N$$

$$b_1 = \lambda Q \, a_N.$$

The medium above the surface is assumed semi-infinite; there is therefore an upward component b_0, but no downward component so that $a_0 = 0$ (Figure 45). The seismometer will record the component b_0 which has to be evaluated. If the detonation takes place at the surface, the conditions of continuity can no longer

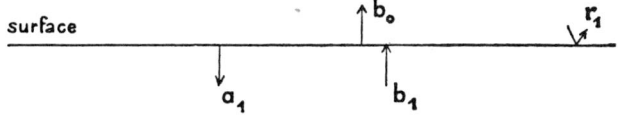

Figure 45. Schematic diagram showing the surface, above which there is assumed to be an infinite half space.

be written as above. It can be said that there is always continuity of the displacements $b_0 = a_1 + b_1$, but that the spectral difference of the stresses on either side of the surface is equal to the spectrum of the source which is equal to 1 for all frequencies. This is expressed by the relation:

$$\rho_0 V_0 b_0 - \rho_1 V_1 (-a_1 + b_1) = 1.$$

We then have the system:

$$b_0 = a_1 + b_1$$

$$\rho_0 V_0 b_0 - \rho_1 V_1 (-a_1 + b_1) = 1,$$

or

$$b_0 = (P + Q) \lambda a_N$$

$$\rho_0 V_0 b_0 - \rho_1 V_1 (-P + Q) \lambda a_N = 1.$$

From it we deduce

$$\rho_0 V_0 b_0 + \rho_1 V_1 \frac{P-Q}{P+Q} b_0 = 1,$$

whence

$$b_0 = \frac{1}{\rho_0 V_0 + \rho_1 V_1} \cdot \frac{P+Q}{P+r_1 Q} \quad \text{with } r_1 = \frac{\rho_0 V_0 - \rho_1 V_1}{\rho_0 V_0 + \rho_1 V_1}.$$

We can then take, as an expression of the impulse seismogram except for a constant factor, the formula:

$$R = \frac{P+Q}{P+r_1 Q} = \frac{a_1 + b_1}{a_1 + r_1 b_1}.$$

This constant factor is equal to the spectrum of the displacement source equivalent to the stress source whose spectrum is 1.

P and Q being polynomials in z, we have a development of the form:

$$R = 1 + c_1 z + c_2 z^2 + c_3 z^3 + \dots,$$

c_i being solely a function of the reflection coefficients. Inasmuch as the inverse transform of z^i is $\delta (t - 2i\tau)$, the transform of R is equal to $\sum_i c_i \delta (t - 2i\tau)$.

In order to obtain a synthetic seismogram corresponding to any input signal, it is sufficient to take the convolution product of the impulse seismogram and the input signal.

Until now we have assumed that source and detector are situated at the surface. More general formulas have been established for the source position at any interface n_1, and that of the detector at an interface n_2. We shall limit ourselves here to indicating that there is no longer continuity across the interface n_1, and that, as a result, the matrix relation can no longer be retained between the beds located on either side of this interface. The sequence of beds is cut into two parts;

we continue to designate the displacement components by a_m and b_m for interfaces with subscript greater than or equal to n_1, and the displacement components with subscript less than or equal to n_1 by $\bar{a}_m$ and $\bar{b}_m$, in keeping with the diagram of Figure 46.

Under these conditions, if the largest and smallest of the quantities n_1 and n_2 are designated by n and q, we have

$$R = \frac{(a_n + b_n)(\bar{a}_q + \bar{b}_q)}{(a_1 + r_1 b_1)\,\bar{a}_1} \prod_{i=2}^{n_1} \frac{1 - r_{i-1}}{1 + r_i},$$

with

$$\prod_2^{n_1} = 1 \quad \text{if } n_1 = 1.$$

Inasmuch as a_1, b_1, a_n, and b_n may be expressed as functions of the reflection coefficients, the delays $z = e^{-2j\omega\tau}$ and a_N, and since $\bar{a}_1$, $\bar{a}_q$, and $\bar{b}_q$ are expressed as functions of the reflection coefficients, the delays and $\bar{a}_0$, we see that R is a function only of the reflection coefficients and the delays.

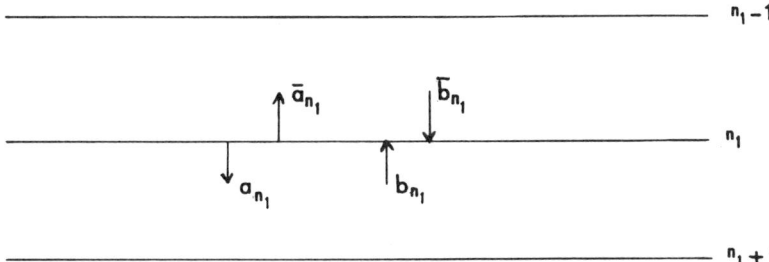

Figure 46. Schematic diagram showing the components $\bar{a}_{n_1}$ and $\bar{b}_{n_1}$ of displacement for the interfaces with index greater than n_1, and the components $a_{n_1} + b_{n_1}$ for interfaces with index of less than n_1.

In the following discussions, we shall refer only to the formula obtained in the case where source and detector are at the surface. We shall see how practical formulas can be inferred in simple fashion for the case of shallow bed filtering; a whole series of linear filters is thus obtained. Noteworthy properties of impulse seismograms with multiples will likewise be inferred from them, properties which allow for implementation of nonlinear demultiplication filtering.

LINEAR INVERSE FILTERS

Filtering of shallow beds

Single-layer case

The impulse seismogram spectrum corresponding to a single layer (Figure 47) has the expression:

$$R = \frac{1 + r_2 z}{1 + r_1 r_2 z} = 1 + (1 - r_1) \frac{r_2 z}{1 + r_1 r_2 z}, \text{ with } z = e^{-2j\omega\tau_1}.$$

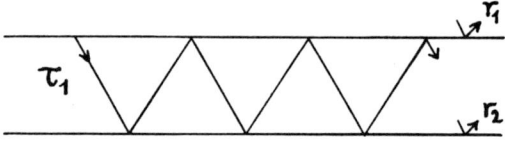

Figure 47. Multiples in a single bed.

If the source and surface transmission are neglected, we have

$$R' = \frac{r_2 z}{1 + r_1 r_2 z} = r_2 z - r_1 r_2^2 z^2 + r_1^2 r_2^3 z^3 \dots \; .$$

It is seen that in addition to the simple reflection $r_2 z$, all the multiple reflections are obtained, displaced in time. In order to preserve only the simple reflection, it is sufficient to multiply R' by the linear factor $1 + r_1 r_2 z$. The resonance due to a surface layer will be generated simultaneously by the source and by each return of a deep reflection, there will therefore be an advantage in applying the filtering twice, hence the expression: $(1 + r_1 r_2 z)^2$. We again encounter the formula given by Backus (1959).

Two-layer case (Figure 48)

The spectrum of the impulse seismogram corresponding to two layers (Figure 48) has the expression:

$$R = \frac{1 + r_2 r_3 z_2 + r_2 z_1 + r_3 z_1 z_2}{1 + r_2 r_3 z_2 + r_1 r_2 z_1 + r_1 r_3 z_1 z_2},$$

with

$$z_1 = e^{-2j\omega\tau_1} \quad \text{and} \quad z_2 = e^{-2j\omega\tau_2}.$$

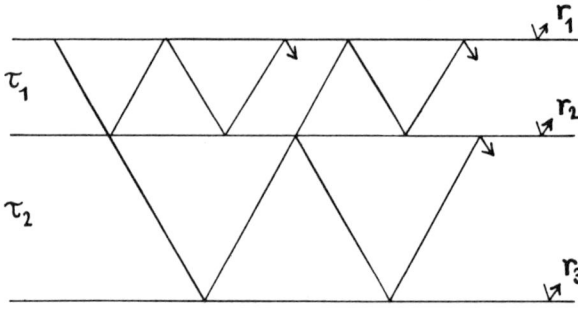

Figure 48. Multiples in two beds.

The denominator of R', raised to the second power, here again constitutes the appropriate linear filter.

Effectiveness of these filters

The formulas given are theoretically interesting, but it must be admitted that their practical application meets with substantial difficulties.

It is indeed known that in the case of a single resonant layer, determination of parameters r_1, r_2, and τ_1 is already awkward since the value of the product $r_1 r_2$ to be taken into consideration on the one hand is not the true value, but the apparent value on the film, while on the other hand the accuracy of trace picking is inadequate in order to determine the value of τ_1. We begin in such a case by evaluating, principally by means of the autocorrelation, an approximate value of τ_1; then by trial and error, an optimal value is fitted.

Experience shows that autocorrelation is no longer of very great help in the presence of several strong reflectors. Incidentally, it is particularly difficult to record on the films all the action of multiple reflections returning to the surface and particularly those of internal multiples. Determination of the number of reflectors i, the value of parameters r_i and τ_i is, furthermore, practically impossible. The difficulty in this determination is especially serious since we are aware of the extreme accuracy required in the choice of τ_i (see Chapter 7).

This filtering has been attempted on a marine seismic section where it is obvious the water bottom is followed by other reflectors which complicate the resonance. This attempt was completely unsuccessful whereas we could think of establishing the filter elements without too much trouble.

In order to study the problem better, we have produced a synthetic seismogram from a continuous velocity log, such that the internal multiples have practically no influence (Figure 49, trace 2). We have artificially created two reflectors at the one-way times $\tau_1 = 24$ msec and $\tau_2 = 104$ msec with reflection coefficients of 0.5 and -0.5, respectively. Synthetic seismograms were obtained with and without multiples (Figure 49, traces 1 and 3). The trace with multiples has been filtered through antiresonance operators with the exact reflection coefficient values in all cases, but allowing for errors in τ_1 and τ_2 the following points can be made.

$\tau_1 = 24$ *msec,* $\tau_2 = 104$ *msec* (*Figure* 49, *trace* 4) — Comparison of traces 2 and 4 shows that the result is, on the whole, satisfactory. The presence of amplitudes which are a little too strong is, however, noted at the two-way times: 452, 524, 820, and 874 msec. Nevertheless they remain weaker than those of the reflection at 714 msec.

$\tau_1 = 23$ *msec,* $\tau_2 = 104$ *msec* (*trace* 5) — The amplitudes displayed in the previous case have assumed far more importance, as much as that of the reflection at 714 msec. There is nothing to indicate, however, that these are multiple reflections of trace 3, which have not been eliminated.

$\tau_1 = 25$ *msec,* $\tau_2 = 106$ *msec* (*trace* 6) — Appearance of cycling about the reflection at 314 msec and of a multiple determined at 758 msec, from the reflection at 710 msec. On the other hand, the multiples at 820 and 874 msec have almost disappeared.

$\tau_1 = 25\ msec$, $\tau_2 = 107\ msec$ *(trace 7)* — The result of filtering has a more unsightly aspect than the seismogram we would want to filter, because of a rather impassive cycling. In this case, not only has no resonance been suppressed, but new ones have been introduced.

It behooves us then to be extremely careful on the effectiveness of approximate antiresonance operators when there are several resonant beds. The margin of error is so small that in practice it is impossible not to exceed it and it is seen that the results of filtering risk being worse than the original recordings.

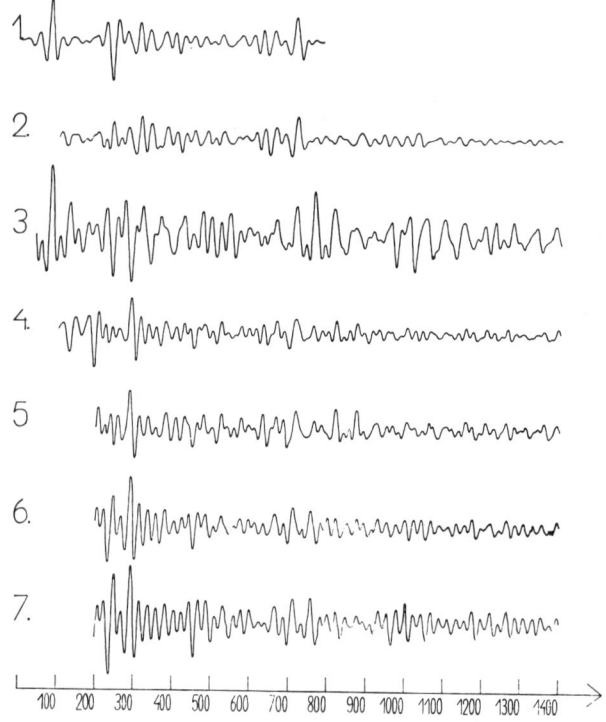

Figure 49. Comparison of synthetic seismograms with minor changes in parameters τ_1 and τ_2 of antiresonance operator. Two-way time is expressed in msec.

Deconvolution

It is known that the seismic impulse resulting from the explosion and transmitted in the earth favors a select band of frequencies. The purpose of deconvolution is to attempt to give the same importance to all frequencies represented in the impulse, over the broadest possible bandwidth. In other words, we wish to suppress the effect of the filtering to which the impulse seismogram has been subjected because of the pulse shape. We arrive here theoretically by producing a contraction of the impulse, as compressed as possible, and a mathematical operator is determined by calculation to perform this contraction.

If the real seismogram is designated by $y(t)$, the impulse by $s(t)$ and the impulse seismogram by $r(t)$, we have, in the absence of noise:

$$y(t) = r(t) * s(t),$$

the star indicating convolution. If we wish to obtain $r(t)$, we must perform the operation:

$$r(t) = \frac{y(t)}{s(t)} = y(t) * \frac{1}{s(t)},$$

the sign ------- indicates deconvolution. The operator sought should at best approach the expression $\frac{1}{s(t)}$.

First, procedures will be described which are sufficiently straightforward for convenient use with the aid of analog devices. It is readily evident that they can also be employed through numerical computation but, because of the extremely simple assumptions which govern their determination, their effectiveness remains very much lower than those calculated by the methods studied in the following section. These operators are too complex to be used in an analog playback center. On the other hand, they are likely to have far higher accuracy and flexibility. We shall not develop the theory of determining deconvolution operators by the least-squares method. These commonly used and effective methods have been adequately reported in the literature, so that it would be pointless to return to the topic in this outline.

Analog methods

The practical realization of the mathematical deconvolution operator is achieved by means of a delay line, whose description has been given (Chapter 4). This equipment allows for a plurality of simultaneous seismic trace readings, the read heads being displaceable in time, with respect to each other and weighted differently. The summation of simultaneous readings is made at the output. The equipment can therefore perform convolution of a trace by a sampled operator.

Properties required for a rapid deconvolution method — A method of seismic pulse compression by deconvolution should possess several properties if we wish to have convenient implementation and it can be readily carried out in the seismic data centers.

Deconvolution being an inverse filter, it is important to determine, perhaps for each trace or portion thereof, or for an entire section, which operator should be applied; this will be discussed later.

Once having established the operator, it should be capable of rapid use. In other words, this operator should be computed in a previously sampled form in a manner requiring only the storage of the coefficients on a delay line. According to the type of delay line available, the inverse operator may be limited in duration and number of samples, or both. The operator should not be too long, nor too complex, so that it may be utilized. Taking the inverse problem we should also be able to say that the equipment to be obtained for implementing deconvolution must be capable of using effective operators. We shall see later what the orders of magnitude are for the elapsed time and the number of necessary samples which govern the type of equipment to be employed.

Finally, while possessing the properties we have just mentioned, the method employed should give rise to a theoretically maximum contraction compatible with the frequency band, in which provision is made to take account of the noise. It is also necessary, and this is contradictory with the last requirement, for the output to be relatively stable — i.e., that minor variations in the spectral properties of the traces have only small influence on the character of the result. It will then be possible to handle a trace, or several traces, with the same operator, or several traces with the one variable operator, the time variant relation being the same. Often a compromise has to be found between desirable properties which act in an opposing sense, the criterion in the final analysis takes account of the filtering effectiveness and its cost.

Methods employed —

a. Inversion of a Ricker-type impulse — The essential assumption is that, in the autocorrelation of a seismic trace, it is principally the abscissas of the first zero and, to a lesser degree, of the first minimum which are representative of the impulse, since the amplitude of the first minimum is often strongly disturbed by the impulse seismogram. It is further assumed that the impulse is symmetrical and of the Ricker type. As this impulse has an analytical expression, it is possible to compute its autocorrelation. A relation is thus established between the width t_0 or t_m separating the first zeros or the first symmetrical minima of the autocorrelation and the center frequency of the impulse spectrum:

$$f_0 \simeq 0.47 \; / \; t_0 \simeq 0.86/t_m.$$

Having completely determined the spectrum of the assumed Ricker impulse $(f/f_0)^2 \, e^{-(f/f_0)^2}$, through the autocorrelation, its inverse is computed, and from it an approximation is established by a cosine series development of the form:

$$\sum_{i=0}^{N} b_i \cos 2\pi f i \tau.$$

The coefficients b_i are determined once and for all. This has been accomplished in such a way that the approximation is valid up to a maximum frequency: $f_m = 3.5 \; f_0 = 1 \, / \, 2\tau$ (f_0 being the center frequency of the impulse spectrum and τ the lag between the coefficients b_i).

This restriction arises from the limitation in the number of coefficients that can be displayed on the delay line. We notice, however, that curve B in Figure 50, which represents the inverse of the Ricker impulse spectrum tends to infinity asymptotically as f tends to zero, and exponentially for large values of f. On the other hand, the maximum attenuation permitted by the delay line, the quality of the approximation we wish to attain in a given frequency band, makes it necessary to impose an upper limit on the values of the Ricker inverse spectrum for the low and high values of frequency. The approximation obtained will be correspondingly better but narrower in a frequency band, as this upper limit weakens. It is therefore necessary to choose a compromise in such a manner that the approximation remains satisfactory for a sufficiently wide frequency band.

In the example shown (Figure 50), the upper limit is equal to one hundred times the smallest value, giving therefore a 40 db attenuation. By varying the argument $2\pi f\tau$ from 0 to π, for a frequency change from 0 to f_m, values of the Ricker impulse inverse spectrum are determined corresponding to: $\dfrac{2i+1}{N+1}\dfrac{\pi}{2}$

($N+1$ being the total number of coefficients to be determined.)

We are aware that curve C representing the variations in $\displaystyle\sum_{i=0}^{N} b_i \cos 2\pi f i\tau$ actually passes through the previously indicated points and that between these points it entwines curve B with values that are sometimes higher and sometimes lower, with the noteworthy property of making the maximum deviation between these curves as small as possible (Angot, 1957, p. 739). Curves B and C are plotted on half scale relative to curves A and D. Curve D represents the product of the Ricker impulse spectrum (curve A) with the cosine series (curve C). It is seen that the deviation is, on the average, from 4–5 percent in the wide band extending from 10 to 95 hz. For values smaller than 10 hz and greater than 95 hz, curve D evidently tends towards zero as a consequence of the upper limit imposed on the inverse spectrum of the impulse.

By varying the upper limit and the peak frequency we establish a suite of filters whose impulse response is

$$b_0\, \delta\,(t) + \frac{1}{2} \sum_{i=1}^{N} b_i \delta\,(t \pm i\tau).$$

In practice, it is therefore sufficient to know which filter is to be applied to determine:
1. the value of the dominant frequency f_0 in the traces to be deconvolved,
2. the value of the maximum attenuation of the filter which depends especially on the dynamic range of the field recording. The actual suite allows us to use filters calculated for f_0 varying from 20 to 40 hz in 2-hz stages, with attenuations of 15, 20, 25, 30, and 35 db.

b. Inverse calculation by polynomial approximation — In this method, the autocorrelation $a(t)$ of the impulse to be contracted is assumed to be known, and

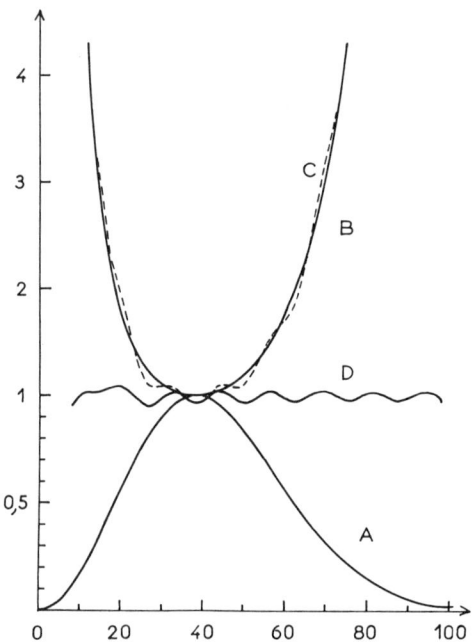

Figure 50.   The spectrum (A) of a Ricker wavelet together with the exact inverse spectrum (B), a truncated approximation (C) and the product (D) of the spectrum and the approximation.

an operator is calculated whose transfer function is approximately the inverse of the square root of the $a(t)$ spectrum. In order for such a procedure to be effective, it is necessary of course that the impulse to be contracted should have an almost real spectrum, that is, the impulse should approach being symmetrical. In our present stage of ignorance as to the form of the impulse, we make the assumption that the phase of its spectrum is zero and a certain contraction is obtained, which indicates that in nature impulses are seldom asymmetrical.

The method that we have used, which is portrayed in Figure 51, 52, and 53, is the method of polynomial approximation. Given that the autocorrelation $a(t)$ has a real spectrum, we can operate on this spectrum with the help of known approximations for the real numbers. We can, for example, find a polynomial $\sigma = c_0 + c_1 X + c_2 X^2$ which may be a good approximation of $1/\sqrt{X}$ in the interval (0.1; 1.1) (see Figure 51). Coefficients c_0, c_1, and c_2 are determined by least squares once and for all, X can be any real number and in particular the spectrum of $a(t)$. If this is so, whatever the frequency may be and provided that the spectrum of $a(t)$ is contained between 0.1 and 1.1, σ will be a good approximation of the operator sought. Let us recall that all these operations may be performed on functions of time, for if X is the spectrum of a, X^2 is that of its autocorrelation. The only condition to be observed is to bring the spectrum of a between 0.1 and 1.1. In general, such is clearly not the case;

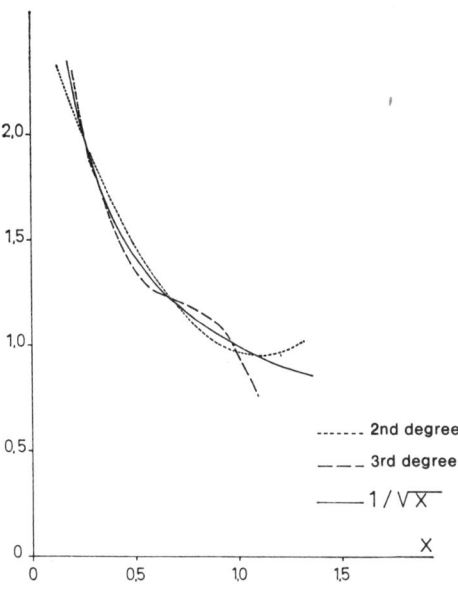

Figure 51.   Polynomial approximations to an inverse spectrum.

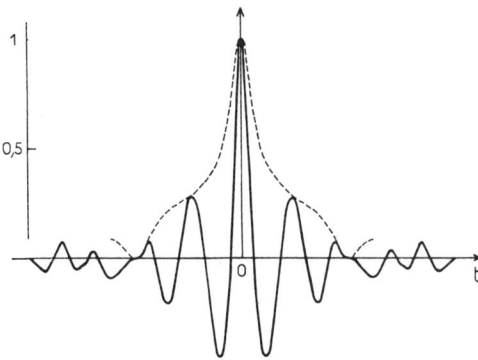

Figure 52.   An autocorrelation in the form of a modulated sine wave with smooth envelope.

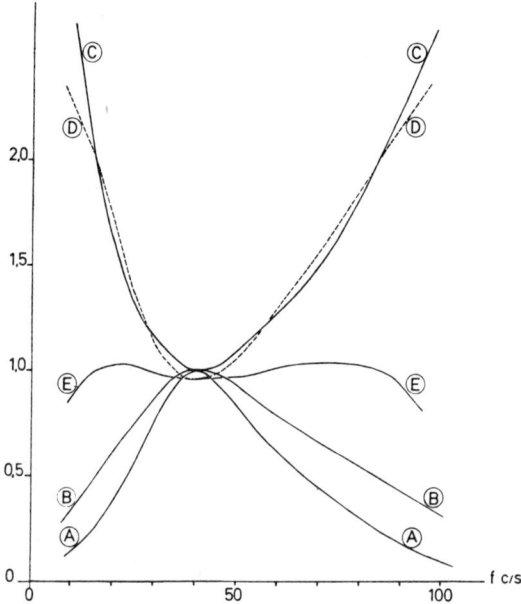

Figure 53.   The spectrum (A) of a normalized autocorrelation with its square root (B). Curve C is the inverse of curve B and curve D is the third degree approximation. Curve E is the spectrum of product of the approximation by the wavelet.

this would only be by reason of the arbitrary scales chosen for measuring the amplitudes of $a(t)$. It is therefore necessary to normalize the autocorrelation by its spectral maximum. The assessment of this value is carried out easily in the majority of cases. In fact, if the autocorrelation $a(t)$ has the form of a sinusoid modulated by a rather smooth envelope (Figure 52), it is easy to show that the spectral maximum is equal to $(T_0/4)\cdot(\Sigma\text{peaks})$, T_0 being the pseudo-period of the sinusoid and (Σpeaks) being the sum of the absolute values of the peaks of $a(t)$. The units for measuring the peak amplitudes are arbitrary provided that the same units are used in the calculation of σ.

Another procedure permits determination of the normalized value in an automatic and satisfactory way by accurately using calculations necessary in the polynomial approximation method. If the autocorrelation of a function $g(t)$ is designated by $a(t)$, we have, according to the Schwarz inequality,

$$\int_{-\infty}^{+\infty} |a(\zeta)|\, d\zeta = \int_{-\infty}^{+\infty} \left|\int_{-\infty}^{+\infty} g(t)\,g(t+\zeta)\, dt\right| d\zeta$$

$$\leqslant \int_{-\infty}^{+\infty} \sqrt{\int_{-\infty}^{+\infty} [g(t)]^2\, dt \int_{-\infty}^{+\infty} [g(t+\zeta)]^2\, dt}\; . \; d\zeta,$$

from which

$$\int_{-\infty}^{+\infty} |a(\zeta)| d\zeta \leqslant \int_{-\infty}^{+\infty} \left[\int_{-\infty}^{+\infty} [g(t+\zeta)]^2 \, dt \right] d\zeta.$$

If the functions are sampled, we shall have

$$\sum_n |a'(n\tau)| \leqslant \sum_m [g(m\tau)]^2 = \sum_m g^2(m\tau),$$

with

$$\sum_n a'(n\tau) \simeq \frac{1}{\tau} \sum_n a(n\tau).$$

Assuming that the function $g(m\tau)$ is a sampled autocorrelation $a_1(m\tau)$, for which we seek to evaluate the spectral maximum, say N, we set

$$a_i(m\tau) = a_{i-1}(m\tau) * a_1(m\tau).$$

We clearly have

$$N^i < \sum_m |a_i(m\tau)|.$$

On the other hand, according to the Schwarz inequality,

$$\sum_m |a_i(m\tau)| \leqslant \sqrt{\sum_{n_1} a_{i-1}^2(n_1\tau). \sum_{n_2} a_1^2(n_2\tau)}.$$

We infer from this that the numbers $y_i = \sum_m |a_i(m\tau)|$ constitute a convergent series in which each term is an upper limit of N^i. By representing the variations in y_i as a function of i, in rectangular coordinates, we can evaluate the minimum of y_i, that is to say an upper limit of N.

Considering an equilateral hyperbola whose asymptotes are parallel to the coordinate axes and which passes through the three points

$$A \begin{vmatrix} 1 \\ y_1 \end{vmatrix}, \quad B \begin{vmatrix} 2 \\ y_2 \end{vmatrix}, \quad \text{and} \quad C \begin{vmatrix} 3 \\ y_3 \end{vmatrix},$$

we can take the ordinate of the asymptote parallel to the i axis as the value of N. We have

$$N \simeq \frac{y_1(y_3 - y_2) + y_3(y_1 - y_2)}{(y_1 - y_2) + (y_3 - y_2)}.$$

In the zone where the variation is practically linear, we can employ a geometric series. For example,

$$N \simeq y_2 + (y_3 - y_2) \, \frac{1}{1 - \dfrac{y_4 - y_3}{y_3 - y_2}} \, .$$

The error in the norm is usually less than 10 percent. It is wise to use a formula valid up to $X = 1.1$, and not just 1, as a precaution in case of error. A third-degree formula is clearly more flexible than one of the second degree.

In summing up, the operations to be performed are the following:

1) compute $a(t)$;
2) take its norm;
3) normalize $a(t)$;
4) calculate σ.

These operations may be carried out on continuous or sampled $a(t)$. If the same sampling interval is used from start to finish in computation, there is no need to take account of its value. If the interval is changed, it becomes necessary to apply a correction which is a function of the ratio of the sampling intervals.

Protection against random noise is automatic. In fact, it is in the frequency band where the impulse spectrum has a low absolute value that we run the risk, when multiplying by the inverse, of building up the random noise in such a way that it approaches or exceeds the level of useful signal. It is therefore necessary to limit the absolute value of the inverse operator transfer function in these bands. In the procedure just described, this is produced automatically since the approximation is being done by default in the region $0 < X < 0.1$. The product of the impulse spectrum with that of the inverse goes to zero at the same time as that of the impulse, as can be seen in Figure 53. The autocorrelation spectrum is given by curve A which is assumed to be normalized. The square root of this spectrum is represented by curve B. Curve C is the exact inverse of curve B. With the help of a third-degree approximation formula (Figure 51), an approximate inverse is obtained which, as can be ascertained from the figure, saturates and remains below the ordinate at the beginning of the curve $1/\sqrt{X}$. The product of the approximate inverse and the impulse (allowing, as we have done at the outset, that this may be symmetrical) has a spectrum which is represented by curve E. It is seen that if the sampling interval is properly chosen, the bandwidth recovered by the deconvolution depends only on the original A spectrum. The assurance of A is at best always fulfilled therefore. Note that if the noise level is less than one tenth the amplitude of the spectral peak, we can take a formula whose approximation falls lower, for example, to 0.05. This is in any case somewhat illusory since formulas of the second- or third-degree are by nature rather approximate, the error in approximation being able to reach 20 to 25 percent in certain places of the useful bandwidth. Experience shows that this is not very troublesome for the error affects only a rather narrow band of frequencies. However, if we wish curve D to give a better mold of curve C in the realm of approximation carried, it would be necessary to choose a formula of a higher degree. In order to improve the inversion while employing a formula of low degree, we can take an approximation which may be constantly deficient. Curve D will then always be below curve C and the rejected frequency band will be broader than it would have needed to be.

The method is deficient if the function employed is not a true autocorrelation. This can happen following errors or approximation in the calculation of $a(t)$ or following a poorly performed truncation of the computed autocorrelation. In this case, the spectrum of $a(t)$ remains quite real if the pseudoautocorrelation employed is symmetrical, but it can become negative in certain regions. As the approximation is valid only for positive values of X, contained for example between 0.1 and 1.1, the spectrum of the approximation takes on aberrant values when X is negative. There is therefore need to take care to use only true autocorrelations. Computation of $a(t)$ should accordingly be carefully made and discretion used in the truncation. It is indeed necessary to truncate the autocorrelation, since only its central portion is used. If the truncation is too abrupt, negative values are introduced into the spectrum. Quite often the autocorrelation envelope calculated from a seismic trace continually approaches the axis, as it departs from the center, then rebounds. We can then truncate at the narrowing of the envelope. In other cases when the trace oscillates appreciably, the autocorrelation has a very flat envelope and application of the method is difficult. In a controversial case where the envelope would not have been well defined, we can multiply the computed autocorrelation by a bell curve having its axis of symmetry coincident with that of the autocorrelation. We can thus expedite the decline of the envelope from the axis of symmetry, which simplifies the decision on where to truncate. The effect that this operation has on the spectrum is to convolve it with a relatively narrow bell curve which rounds it off somewhat but does not introduce negative values.

c. Comments on autocorrelation — With regard to the use of autocorrelation of the trace to extract an approximation of $a(t)$, the impulse autocorrelation, it should be noted that actually we do not know beforehand what the relation is which links them in the absence of complete knowledge of the impulse seismogram. The deconvolution performed is therefore not necessarily a deconvolution by the primary impulse, but a deconvolution by something which depends on the impulse seismogram as well as the impulse itself. Nevertheless, if the autocorrelation is reasonably truncated, retaining only about 60 msec on either side of the axis of symmetry, it is certain that a smooth spectral inversion will be achieved. Oscillations of this spectrum would not have a period less than $1000/60 = 16.7$ hz and it is felt that the majority of information contained therein pertains to the impulse and not the impulse seismogram. The latter, in fact, outside cases of resonance due to the presence of a large reflection coefficient at shallow depth, has statistical properties such that no frequency is especially favored (Agard and Grau, 1961). If we improve its spectrum by truncation in the time domain, we should expect to obtain a rather flat spectrum. It seems then, that once the width correction is made (see above), we can use the trace autocorrelation to perform calculation of the inverse, provided that the latter is not too long. Moreover, we need to know that the end oscillations do not have a large value, which further indicates that it is quite in order to multiply the trace autocorrelation with a Gaussian curve. In fact this modifies only the end oscillations, and those at the center are preserved. In any case, it is readily seen that deconvolution from trace autocorrelation, whatever the procedure used may be, has limited effectiveness due to the fact that we do not even truly know the impulse autocorrelation.

It is, however, striking to ascertain that in employing the methods which have been described and with a minimum of trial and error, we arrive at a contraction which, without reaching a great value, nonetheless utilizes almost the entire frequency band where the spectrum of the autocorrelation is higher than the peak value by at least 20 or 30 db. That we arrive at the anticipated contraction readily indicates how weak the phase effect is.

d. Electrical filtering — Finally a few words about electrical filtering which has given some interesting results in the special case presented. The slope of the filter corresponding to the inverse of a Ricker impulse spectrum is very regular between the central frequency and the high frequencies. An attempt has been made to approximate this slope through a simple electrical filter, which attenuates frequencies below 60 hz, in the present circumstances the K60 filter. We thus attain the very idea of Ricker who advocated an exponential filter for the high frequencies $(e^{\alpha f^2}$ with $\alpha > 0)$ which would broaden the spectrum $(f/f_0)^2 \, e^{-(f/f_0)^2}$ and achieve a contraction of the impulse.

Comparison of Methods — Contradiction exists between the philosophy of the method which employs a suite of inverse Ricker pulses and that of polynomial approximation. The former in effect assumes that the impulse autocorrelation is well defined by the position of the first troughs of the trace autocorrelation. The second method takes into account all the trace autocorrelation events, at least in the central portion.

The practical examples presented here do not seem adequate to decide between the two rival methods which give essentially equivalent results. The polynomial method permits spectral modelling of the inverse operator on any spectrum and for this reason is more flexible. However, this advantage is somewhat deceptive, for it is seldom that autocorrelation events other than the first two troughs, have great meaning. They only have significance when the envelope of the central portion of the autocorrelation is sufficiently broad and smooth. Moreover, in such a case the impulse oscillates more than a Ricker impulse.

In Figure 54, curve A represents, for a special case (peak frequency of 25 hz and 15-db attenuation), the transfer function of the filter calculated by the first method; curve B — that of the filter calculated by the second method (polynomial approximation). We note that the latter more readily passes the 30 — 50 hz band. By contrast, curve A favors the high frequencies from 60 hz onwards.

It is therefore likely that the two methods have different fields of application which partially overlap. The seismic section considered in this article would then be in the common field.

It will be noted that, whether by the method which uses the inverse Ricker impulses, or by polynomial approximations, we are led to rather long operators. They easily reach $300 - 400$ msec. It is essential therefore to make use of a delay line which allows for application of delays of this order without producing appreciable attenuation or phase distortion over a bandwidth up to and including the highest frequencies usable. The equipment that we have employed is capable of

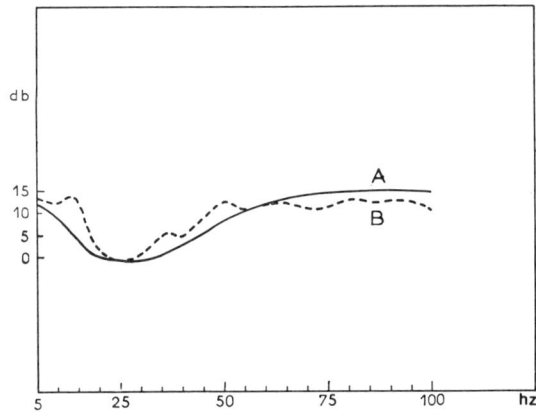

Figure 54. Comparison of the transfer function computed by inverse of the Ricker pulse (A) and by polynomial approximation (B).

giving delays of one second or more without significant distortion up to frequencies of 400 hz (-3 db) and without notably adding to the noise, the dynamic range being from 50 to 60 db. It therefore has characteristics which well surpass the requirements in deconvolution and which permit several successive stages of filtering, for example, deconvolution and fan filtering. The second consequence of the operator length has a bearing on the mode of recording. In order for the pulse contracting filter to have full effectiveness, it is necessary for the time constant of the automatic gain control to be greater than the length of the operator. It is therefore necessary to employ only slow AGC or to record carefully with programmed gain or further, which is the most convenient, to employ both. Recordings used for these tests of pulse contraction on a production basis have been taken with programmed gain and very slow AGC.

Preliminary operations in practical application —

   a. Practical performance of mathematical operators — We start by determining autocorrelations with the optical correlator. In the particular case presented, the autocorrelations obtained are approximate, since they result from a summation of the absolute values of the difference in amplitudes instead of summing the amplitude products; this stems from the fact that we have operated in wide trace instead of operating in variable density. The abscissas of the zeros and extrema are preserved, but the amplitude ratio of these extrema is not. At the center, this ratio is 2 and tends progressively towards 1 as we depart from the center; this ratio is moreover rapidly reached when the amplitudes are weak compared to that of the central peak (see Chapter 5).

The assumption made in the method of the inverse Ricker impulse permits immediate determination of the central frequency of the desired filter. It is adequate therefore to try several types of filtering with different attenuation and to compare the results.

In the second method, the norm (amplitude of the special maximum) remains to be found and the operator computed: $c_0 + c_1 \, a \, (t) + c_2 \, a^{*2} \, (t) + \ldots \, [a^{*2} \, (t)$ denotes the autoconvolution of the autocorrelation $a(t)$]. At the present time these two computations are programmed for the IBM 7040, which requires sampling and truncation prior to autocorrelation, but it seems that because the coefficients c_i are calculated once for all, we can perform the operation by optical correlator.

b. Determination of traces that can be filtered with the same operators — The deconvolution operation only offers an advantage when the recordings are good, a first criterion permitting determination of traces that can be filtered with the same operator is the preservation of reflection character on these traces. Although somewhat subjective, this criterion is excellent, for the same reason as detection of energy arrivals by eye, through visual correlation of amplitude and character of several traces. Since determination of autocorrelations by the optical correlator is economical, a second criterion is the analogy of the autocorrelations of several traces. The stability of abscissas for the first zero and the first minimum will be especially considered, as well as the analogy of autocorrelation envelopes. However that may be, there is no point in over subdividing the zones of application of the one operator, the resulting section risks lacking uniformity. Moreover, considering the operating time involved, it takes hardly longer to filter successively the entire section with different operators and so obtain several sections for interpretation instead of only one.

Presentation of results —

a. Original section (Figure 55a) — The section shown, taken from a profile of 216 traces corresponding to a length of approximately 13 km, comprises 96 traces. The distance between traces is 60 m. The field record was made with a filter eliminating frequencies below 20 hz, programmed gain and very slow AGC. A two-way time scale appears at the side of the section. On the whole, the evidence is good; it is four-fold coverage. We note, however, a few portions of mediocre quality in which deconvolution will not produce much. The horizons have good continuity, especially around 1500 msec where higher-frequency events (35 hz) are to be found. At 1400 msec, two rather blurred loss-frequency events are noted, in the midst of which appear rough outlines of high frequency impulses, corresponding to inflection points in wiggle trace presentation. Diffraction hyperbolas will be seen with crests located close to 2100 msec. Finally, a wedge between 1600 and 1700 msec poses a picking problem of convergence on some phases.

b. F 20 db (Figure 55b) — This section is the result of filtering the previous one by the Ricker inverse centered on 26 hz with a maximum attenuation of 20 db. The remarkable separation of the arrival at 1670 msec will be noted. At 1400 msec there is clear assurance of the character of the energy arrivals. On the original section, this separation effectively took place over three or four traces; one would guess on others. Here it is continuous. A high frequency phase is discernible even on the diffraction hyperbolas which allows for better definition of the crest.

Elsewhere, the reflection continuity can be followed five traces further than on the original section, especailly towards 1250 msec. In this zone the character is preserved.

The triangular picks indicated on the previous profile is duplicated; it is probably a case of interference between reflections and a diffraction pattern. Other tests performed show a progressive tendency toward uniformity as the maximum attenuation is increased: apparent disappearance of the low frequencies, equalization of amplitude and character, increased tendency to resonance.

c. PI 3 D (Figure 56) — This section is obtained with the help of an operator calculated by the method of polynomial expansion. The filter is centered at 25 hz and its attenuation reaches 15 db. The 25 hz seems very subdued with respect to the 35 hz. The similarity of this section with the example of the inverse Ricker impulse procedure is striking. The two sections differ only by this weaker gain and a resolving power very slightly less in the present case. We have the same appearance of high frequencies towards 1500 msec in the diffraction hyperbolas, the same duplication in the triangular convergence.

Numerical procedures

The delay line employed in the analog procedures automatically limited the number of samples of the mathematical deconvolution operators by the very nature of its construction. This inconvenience no longer exists if numerical calculations are employed. The accuracy of three procedures will be given which make use of the entire assumption that the central portion of the trace autocorrelation is representative of the impulse autocorrelation.

Iterative method of determining the impulse inverse — If the actual trace is denoted by $y(t)$, the impulse by $s(t)$, the problem consists of finding $r(t)$, such that

$$r(t) = \frac{y(t)}{s(t)} = y(t) * \frac{1}{s(t)}.$$

It is assumed that the impulse is symmetrical, its autocorrelation $a(t)$ is such that its spectrum $A(f)$ is normalized at 1, by one of the previously indicated methods. Under these conditions we obtain an approximation of: $S'(f) = 1/S(f) = 1/\sqrt{A(f)}$ by applying an iterative formula of Raphson-Newton to solve the equation:

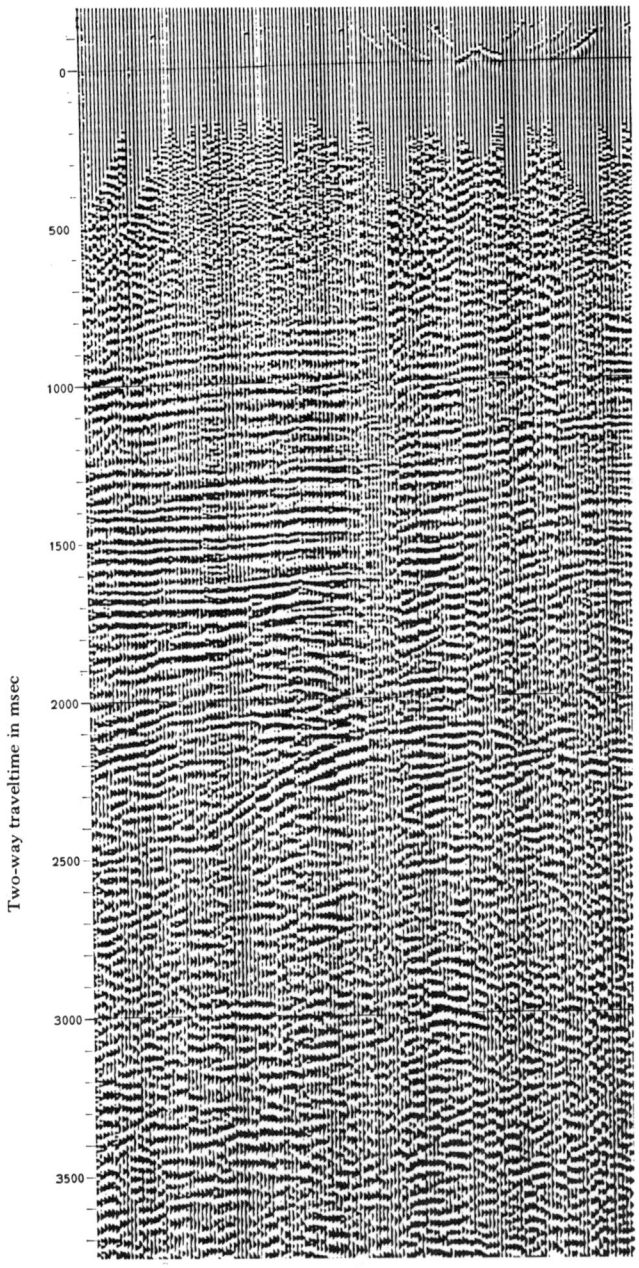

Figure 55a.   Record section 13 km long and including 96 traces.

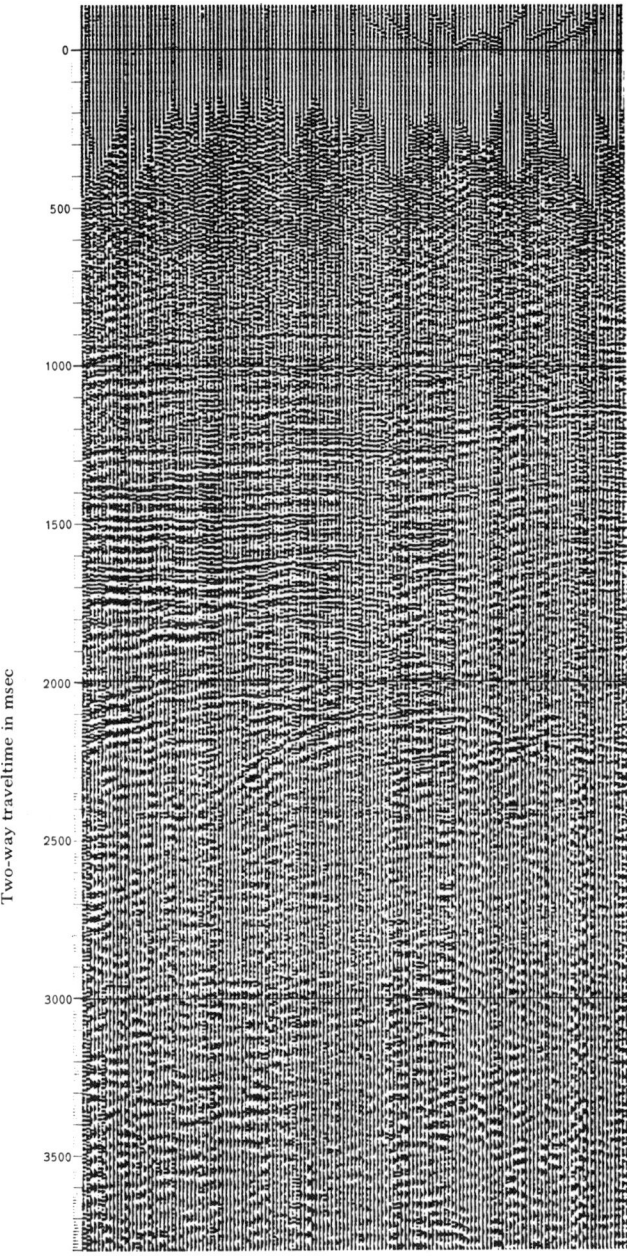

Figure 55b. Record section obtained from (50a) by filtering with the inverse of a Ricker wavelet centered at 26 hz and with a maximum attenuation of 20 db.

Two-way traveltime in msec

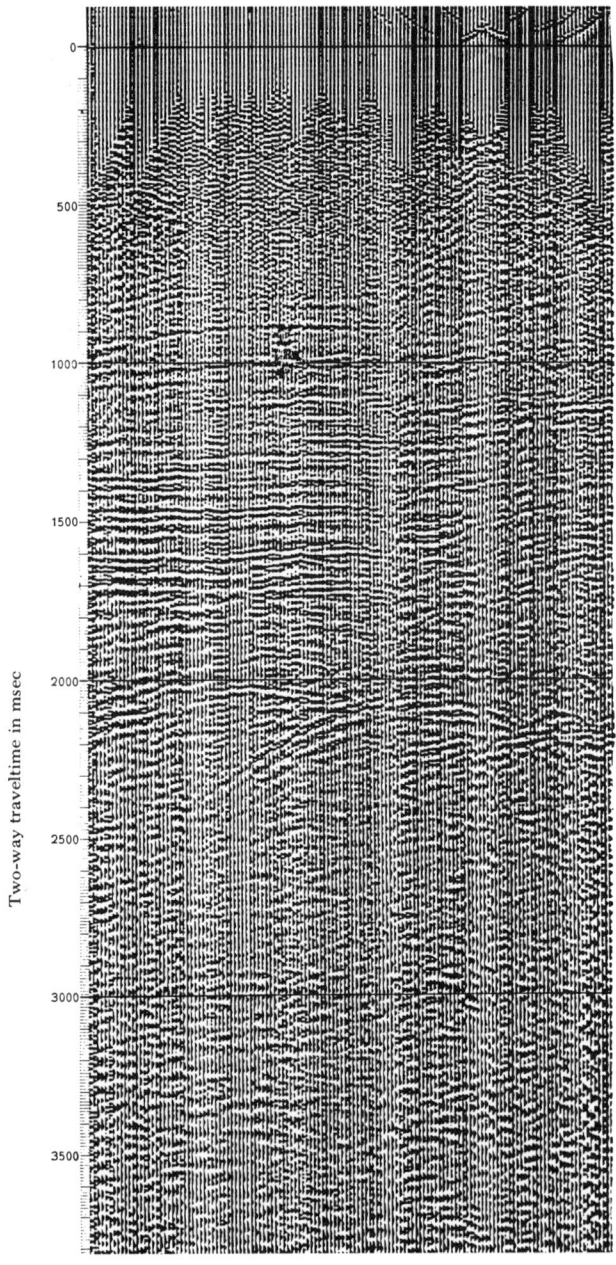

Figure 55c. Record section obtained from (50a) by filtering with operator formed by polynomial approximation, centered at 25 hz, and with 15 db attenuation.

$$1 - \frac{1}{S'^2(f)\,A\,(f)} = 0.$$

We have

$$S_i'\,(f) = S_{i-1}'\,(f)\,\frac{3 - S_{i-1}'^2\,(f)\,A\,(f)}{2}.$$

If this formula is transposed into the time domain, we have

$$s_i'\,(t) = s_{i-1}'\,(t)\,\frac{3 - s_{i-1}'\,(t)\,*\,s_{i-1}'\,(t)\,*\,a\,(t)}{2}.$$

If, in the spectrum $A(f)$ of the autocorrelation, we have taken the precaution to eliminate weak values close to zero or even equal to zero, convergence of the previous formula is rather rapid. It remains only to carry out the convolution of $y(t)$ by the impulse inverse found.

Method of roots — We again assume that the impulse is symmetrical. Its spectrum $S(f)$ is then calculated such that: $S\,(f) = \sqrt{A\,(f)}.$

Application of the inverse Fourier transform produces the impulse $s(t)$, whose inverse is to be calculated. The symmetrical impulse obtained is sampled:

$$s\,(t) = s_0\,\delta\,(t) + s_1\,\delta\,(t + \tau) + s_1\,\delta\,(t - \tau) + \dots + s_N\,\delta\,(t + N\tau) + s_N\,\delta\,(t - N\tau).$$

Its spectrum can be written

$$S\,(f) = s_0 + s_1\,e^{2\pi jf\tau} + s_1\,e^{-2\pi jf\tau} + \dots + s_N\,e^{+2\pi jfN\tau} + s_N\,e^{-2\pi jfN\tau}.$$

If we set $e^{-2\pi jf\tau} = z$, we obtain

$$S\,(z) = s_0 + s_1 z^{-1} + s_1 z + \dots + s_N z^{-N} + s_N z^N$$

or
$$S\,(z) = z^{-N}\,(s_0 z^N + s_1 z^{N-1} + s_1 z^{N+1} + \dots + s_N + s_N z^{2N}).$$

We therefore have a summetrical polynomial of degree 2N. To every root z_i, there corresponds the root $1/z_i^*$. If we have taken the precaution to suppress the weak values of the autocorrelation spectrum $A(f)$, there will be no root with modulus equal to 1. We now group the roots with moduli greater than unity and form the polynomial of which they are the roots. The inverse of such a polynomial is convergent and can easily be calculated numerically. We likewise group the roots with modulus less than unity and form the polynomial with these roots. If we introduce a change in variable $\xi = z^{-1}$ in this polynomial, the coefficients obtained are the same as those of the first polynomial; it will be the same with the inverse. It suffices then, in order to obtain the impulse inverse, to carry out the autocorrelation of the first inverse obtained. A precursor has thus been introduced, whose role is not troublesome, in the convolution of $y(t)$ by the inverse.

Example: Assume that we have obtained the impulse:

$$s\,(t) = 5.2\,\delta\,(t) + \delta\,(t + \tau) + \delta\,(t - \tau).$$

We have successively:

$$S\,(f) = 5.2 + e^{2\pi j f \tau} + e^{-2\pi j f \tau}.$$

$$S\,(z) = 5.2 + z^{-1} + z,$$

$$= z^{-1}\,(1 + 5.2\,z + z^2).$$

The roots of the polynomial $z^2 + 5.2\,z + 1$ are

$$z_1 = -5 \quad \text{and} \quad z_2 = -0.2.$$

The polynomial whose roots in absolute value are greater than 1 is: $1 + 0.2z$, which has the inverse:

$$1 - 0.2\,z + 0.04\,z^2 - 0.008\,z^3 + 0.0016\,z^4 - 0.00032\,z^5 + \dots \ .$$

The polynomial whose roots in absolute value are less than 1 is: $1 + 5z$ or: $5\,z\,(1 + 0.2\,z^{-1})$, which has the inverse:

$$0.2\,z^{-1}\,(1 - 0.2\,z^{-1} + 0.04\,z^{-2} - 0.008\,z^{-3} + 0.0016\,z^{-4} - 0.00032\,z^{-5} + \dots).$$

The inverse of $S(z)$ is therefore:

$$\frac{1}{S\,(z)} = 0.2z^{-1}\,(\dots - 0.00032\,z^{-5} + 0.00166\,z^{-4} - 0.00833\,z^{-3} + 0.04166\,z^{-2}$$

$$- 0.20833\,z^{-1} + 1.04166 - 0.20833\,z + 0.04166\,z^2 - 0.00833\,z^3$$

$$+ 0.00166\,z^4 - 0.00032\,z^5 + \dots),$$

in limiting the coefficients to 5×10^{-6}.

Method of minimum-phase operators — In the case of deconvolution, minimum-phase operators can be calculated in several ways:

We consider $\displaystyle\sum_{i=-N}^{N} a\,(i\tau)$ as being the sampled autocorrelation whose spectrum is $\displaystyle\sum_{i=-N}^{N} a_i\,z^i$ with $z = e^{-2j\omega\tau}$.

By calculating the roots of the symmetrical polynomial $\displaystyle\sum_{i=-N}^{N} a_i\,z^i$ and grouping the roots of moduli greater than 1, a polynomial $\displaystyle\sum_{i=0}^{N} s_i\,z^i$ is formed such that:

$$\sum_{i=0}^{N} s_i z^i \sum_{i=0}^{N} s_i z^{-i} = \sum_{i=-N}^{N} a_i z^i.$$

The corresponding sampled impulse $\displaystyle\sum_{i=0}^{N} s(i\tau)$ is an impulse with minimum

phase. Dividing the actual sampled seismogram $\displaystyle\sum_{i} y(i\tau)$ by this impulse, we ac-

complish deconvolution by the minimum-phase operator.

In Papoulis (1962), relations will be found linking the phase to the modulus in the case of minimum-phase operators.

In practice, we employ another procedure instead based on the fact that every function of $z = e^{-j\omega t}$, whose real part of the spectrum is positive, is stable. Let

us consider, for example, the polynomial $P(z) = \dfrac{1}{1 + \displaystyle\sum_{n=1}^{N} a_n z^n}$, as that occuring

in deconvolution. It is known that such a polynomial will be convergent (or stable) if the roots $z_m = x_m + jy_m$ are located outside the circle of unit radius in the complex plane (x, y).

If we set $P(z) = \dfrac{1}{Q(z)}$, with $Q = \xi + j\eta$, and envisage a conformal repre-

sentation of the complex plane (x, y) of z on the complex plane (ξ, η) of Q, it is seen that "the origin $Q = 0$ $(\xi = 0, \eta = 0)$ of plane Q corresponds to all the zeros of the polynomial Q. The unit circle of the z plane is represented by the curve L in the Q plane," (Brillouin, 1959, p. 273-274) whose parametric equations are none other than $\xi = \xi(\omega), \eta = \eta(\omega)$: the curve L is therefore a representation of the spectrum of polynomial Q. It is evident that if curve L does not encompass the origin of the Q plane, the polynomial Q has no roots located within the circle of unit radius of the z plane. This condition is satisfied, if the real part ξ of Q is assigned to be positive: $\xi(\omega) > 0$. Thus the entire function $Q(z)$ with a real positive part is stable; the converse is not true.

We therefore obtain, in this manner, perfectly stable operators whose inverses are obtained at once without prior calculation of the roots. Impulse seismograms with multiple reflections (nontruncated) possess this property of having a spectrum with a real positive part. The advantage of the method is in having to work only the real part of the spectrum, which constitutes an appreciable time saving. The application to deconvolution is shown to be particularly effective.

We can proceed in the following manner. We begin by calculating the real part of the trace to be deconvolved. A function is then determined whose sampled values are equal to the absolute values of the real part calculated. We next perform a smoothing on the values so obtained. The spectrum smoothed in this manner will be considered as the impulse spectrum. Term-by-term division of real part values by values of the smoothed spectrum give values of the spectrum's real part for the deconvolved trace. The inverse Fourier transform gives a function of time very close to those obtained by deconvolution via a symmetrical impulse with the method of roots, if we restrict ourselves in both cases to the same frequency band. In fact, the procedure indicated here is much more powerful because there is theoretically only the sampling step in time which limits the frequency band. In practice it is appropriate, however, to take account of the recorded noise, especially in the case of analog recording, and not to accept too low a threshold value, which would risk amplifying the noise dangerously.

NONLINEAR FILTERING — THE DEMULTIPLE PROCESS

If we assume that the deconvolution has been sufficiently effective so as to obtain an approximation of the impulse seismogram with multiples, from the actual seismogram, we can attempt to go back to the reflection coefficients, even to the velocity law. With this end in view, it is necessary to unravel the process which permits us to obtain, by matrix products, impulse seismograms with multiples. For this, we must constrain the result of deconvolution to realize a necessary condition, which will be demonstrated here.

It is first verified that the function obtained in effecting the sum of an impulse seismogram and of its symmetrical counterpart, is an autocorrelation function, that is to say, a function whose spectrum is real and positive. It is then a matter of showing that we have

$$R_N + R_N^* = \frac{P_N + Q_N}{P_N + r_1 Q_N} + \left(\frac{P_N + Q_N}{P_N + r_1 Q_N} \right)^* \geqslant 0.$$

Denoting the previous sum by A, we have, successively:

$$A = \frac{(P_N + Q_N)(P_N^* + r_1 Q_N^*) + (P_N^* + Q_N^*)(P_N + r_1 Q_N)}{(P_N + r_1 Q_N)(P_N^* + r_1 Q_N^*)}$$

$$= \frac{(1 + r_1)(P_N + Q_N)(P_N^* + Q_N^*) + (1 - r_1)(P_N P_N^* - Q_N Q_N^*)}{(P_N + r_1 Q_N)(P_N^* + r_1 Q_N^*)}.$$

The denominator being the product of a spectrum with its conjugate, it is necessarily real and positive. It remains to be shown that the numerator is real and positive. The first term is, since $1 + r_1$ is real and positive, and $(P_N + Q_N)(P_N^* + Q_N^*)$ is the product of a spectrum with its conjugate. The factor $(1 - r_1)$ being real and positive, it suffices to show that $P_N P_N^* - Q_N Q_N^*$ is likewise. We demonstrate this by recurrence.

We have

$$P_1 = 1 \quad \text{and} \quad Q_1 = r_2 z,$$

whence

$$P_1 P_1^* - Q_1 Q_1^* = 1 - r_2^2 \text{ which is real and positive.}$$

We assume this property holds for $P_n P_n^* - Q_n Q_n^*$ and show that:

$$P_{n+1} P_{n+1}^* - Q_{n+1} Q_{n+1}^*$$

is likewise real and positive. In the theory of impulse seismograms, we have set

$$\begin{pmatrix} P_n \\ Q_n \end{pmatrix} = \prod_{m=2}^{n+1} \begin{pmatrix} 1 & r_m \\ r_m z & z \end{pmatrix} \begin{pmatrix} 1 \\ 0 \end{pmatrix}.$$

We ascertain that it may be written

$$\begin{pmatrix} P_{n+1} \\ Q_{n+1} \end{pmatrix} = \begin{pmatrix} P_n & z^n Q_n^* \\ Q_n & z^n P_n^* \end{pmatrix} \begin{pmatrix} 1 & r_{n+2} \\ r_{n+2} z & z \end{pmatrix} \begin{pmatrix} 1 \\ 0 \end{pmatrix},$$

since it can be verified that each elementary matrix:

$$\begin{pmatrix} 1 & r_m \\ r_m z & z \end{pmatrix}$$

has the form

$$\begin{pmatrix} P & z Q^* \\ Q & z P^* \end{pmatrix}$$

with $P = 1$ and $Q = r_m z$.

This expression gives the relations:

$$P_{n+1} = P_n + r_{n+2} z^{n+1} Q_n^*$$

$$Q_{n+1} = Q_n + r_{n+2} z^{n+1} P_n^*.$$

From them, we deduce

$$P_{n+1} P_{n+1}^* = P_n P_n^* + r_{n+2} z^{n+1} P_n^* Q_n^* + r_{n+2} z^{-(n+1)} P_n Q_n + r_{n+2}^2 Q_n Q_n^*$$

$$Q_{n+1} Q_{n+1}^* = Q_n Q_n^* + r_{n+2} z^{n+1} P_n^* Q_n^* + r_{n+2} z^{-(n+1)} P_n Q_n + r_{n+2}^2 P_n P_n^*,$$

and finally

$$P_{n+1} P_{n+1}^* - Q_{n+1} Q_{n+1}^* = (1 - r_{n+2}^2)(P_n P_n^* - Q_n Q_n^*) = \prod_{m=2}^{n+2} (1 - r_m^2).$$

The product of the second member is real and positive, and the characteristic mentioned above is verified. We then have

$$R_N + R_N^* \geqslant 0.$$

When the surface coefficient is different from -1, this condition is not sufficient. To find a more general condition, we can express the problem in the following manner. Let there be a continuous time function $r(t)$ with spectrum $R(z)$ and $R(0) = 1$. This function being sampled with an interval τ such that $z = e^{-2j\omega\tau}$, it is a matter of finding which condition the R_n series must satisfy in order that the relation:

$$\frac{P_n + Q_n}{P_n + r_1 Q_n} = R_n$$

permits determination of a suite of reflection coefficients r_i with absolute value less than 1. According to the foregoing, we see that it is necessary for the expression $P_n P_n^* - Q_n Q_n^*$ to be positive or zero, regardless of what n may be.

From the relation $R_n = \dfrac{P_n + Q_n}{P_n + r_1 Q_n}$ we deduce

$$\frac{Q_n}{P_n} = \frac{R_n - 1}{1 - r_1 R_n} = C.$$

The inequality $P_n P_n^* - Q_n Q_n^* \geqslant 0$ is now written

$$P_n P_n^* (1 - CC^*) \geqslant 0, \text{ which reduces to } 1 - CC^* \geqslant 0.$$

If C is expressed as a function of R_n, we obtain

$$1 - \frac{(R_n - 1)(R_n^* - 1)}{(1 - r_1 R_n)(1 - r_1 R_n^*)} \geqslant 0.$$

As the denominator is an autocorrelation, the inequality reduces to

$$(1 - r_1 R_n)(1 - r_1 R_n^*) - (R_n - 1)(R_n^* - 1) \geqslant 0,$$

which can be written, all calculations made:

$$R_n + R_n^* - (1 + r_1) R_n R_n^* \geqslant 0.$$

In order to apply this condition, we encounter serious difficulties. Without entering into details here, let it suffice to indicate that we have obtained good results in assigning the real part of R_n to be greater than $(1 + r_1)/2$ and less than $2/(1 + r_1)$. Justification for the last condition is found easily enough, but not for the first.

The region of validity for the value of the real part can be graphically portrayed as a function of r_1 (Figure 56).

It is observed that the two curves are inverse to each other with respect to the constant 1.

Figure 57 displays, in part 1, a function whose first terms are unknown, as often happens on seismic recordings. In part 2, a function obeying the general

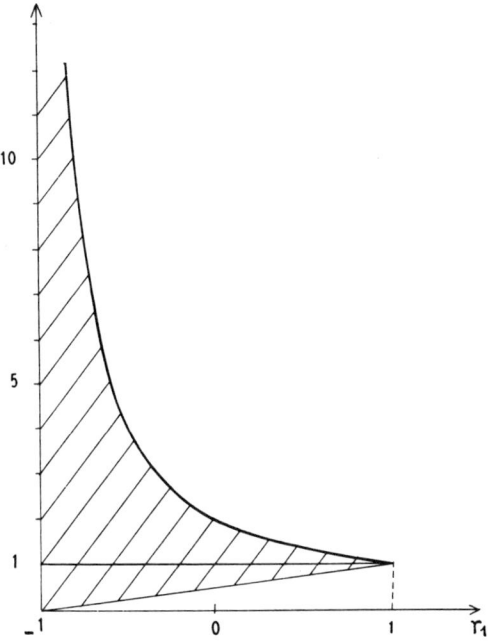

Figure 56.   Domain of validity of the real part of R_n as a function r_1.

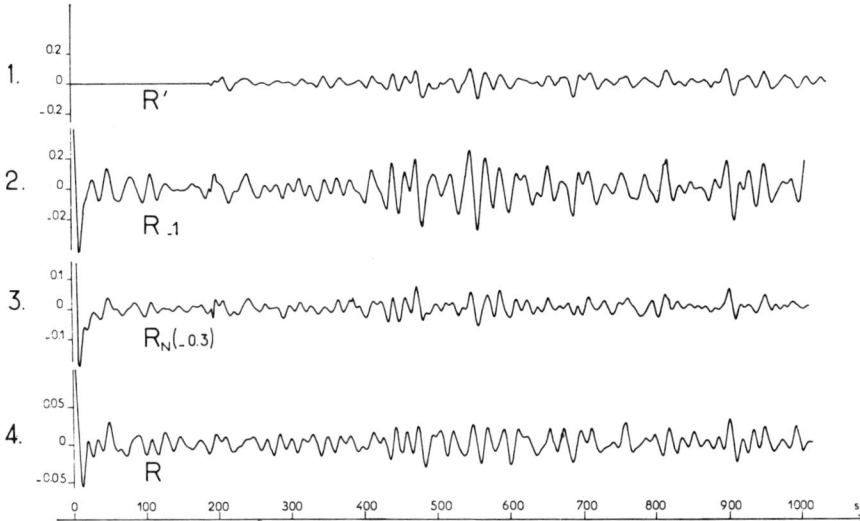

Figure 57.   Synthetic seismograms under assumption of 1) unknown near-surface conditions; 2) surface reflection coefficient equal to —1; 3) and 4) surface reflection coefficient equal to — 0.3, but different computation methods.

condition that r_1 equals -1 is shown. In parts 3 and 4 are some functions obtained by different procedures and conforming to the general condition when r_1 equals -0.3. When it is assured that the function $r(t)$ is such that its spectrum $R(z)$, with $R(0) = 1$, conforms to the previous condition, we can conveniently calculate the suite of r_i, of the reflection coefficients. We have

$$R_n = \frac{P_n + Q_n}{P_n + r_1 Q_n} = 1 + c_1 z + c_2 z^2 + c_3 z^3 + \dots + c_m z^m + \dots$$

$$= 1 + (1 - r_1)\, r_2 z + (a_2 + b_2)\, z^2 + (a_3 + b_3)\, z^3 + \dots + (a_m + b_m)\, z^m + \dots$$

where a_m = the multiple contribution of the first $m - 1$ beds, and
b_m = the contribution of the simple reflection of the mth bed:

$$b_m = r_{m+1}\, (1 - r_1) \prod_{i=2}^{m} (1 - r_i^2) \text{ with } m \geqslant 2.$$

The coefficients c_m are known; then by identification:

$$(1 - r_1)\, r_2 = c_1$$

whence the value of r_2,

$$a_3 + b_3 = c_3$$

$$\dots\dots\dots\dots$$

$$a_m + b_m = c_m.$$

$$\dots\dots\dots\dots$$

If we assume the r_i reflection coefficients to be known for $2 \leqslant i \leqslant m$, we can calculate the impulse seismogram with multiples for the first $m - 1$ beds, having determined the polynomials P_{m-1} and Q_{m-1}.

We have
$$\binom{P_{m-1}}{Q_{m-1}} = \prod_{i=2}^{m} \begin{pmatrix} 1 & r_i \\ r_i z & z \end{pmatrix} \binom{1}{0}$$

and
$$R_{m-1} = \frac{P_{m-1} + Q_{m-1}}{P_{m-1} + r_1 Q_{m-1}} = 1 + d_1 z + d_2 z^2 + d_3 z^3 + \dots + d_m z^m + \dots.$$

There is evident identity between the coefficients c_i and the coefficients d_i for $1 \leqslant i \leqslant m - 1$. Next, the coefficient d_m gives the multiple contribution of the first $m - 1$ layers at the time $2\, m\tau$, such that $z^m = e^{-2j\omega m\tau}$.

We then have
$$a_m = d_m,$$

and thence
$$c_m = d_m + b_m = d_m + r_{m+1}\, (1 - r_1) \prod_{i=2}^{m} (1 - r_i^2).$$

This equation determines r_{m+1}.
For example, for $m = 2$, we have

$$r_2 = \frac{c_1}{1 - r_1},$$

from which we deduce

$$\begin{pmatrix} P_1 \\ Q_1 \end{pmatrix} = \begin{pmatrix} 1 & r_2 \\ r_2 z & z \end{pmatrix} \begin{pmatrix} 1 \\ 0 \end{pmatrix},$$

whence

$$P_1 = 1 \quad \text{and} \quad Q_1 = r_2 z,$$

and

$$R_1 = \frac{P_1 + Q_1}{P_1 + r_1 Q_1} = 1 + d_1 z + d_2 z + ...$$

with $d_1 = r_2 (1 - r_1)$, hence the identity with c_1.

$d_2 = - r_1 r_2^2 (1 - r_1)$, which is the multiple contribution at the instant 4τ. Thence we calculate r_3 by the equation:

$$r_3 (1 - r_1) (1 - r_1^2) - r_1 r_2^2 (1 - r_1) = c_2.$$

For $m = 3$ we calculate P_2 and Q_2 by the recurrence formulas:

$$P_{m-1} = P_{m-2} + r_m z^{m-1} Q_{m-2}^*$$
$$Q_{m-1} = Q_{m-2} + r_m z^{m-1} P_{m-2}^*.$$

From it, we deduce R_2 and d_3,
whence

$$r_4 (1 - r_1) (1 - r_2^2) (1 - r_3^2) + d_3 = c_3, \text{ etc.}$$

In fact, in the division of polynomials $R_m = \dfrac{P_m + Q_m}{P_m + r_1 Q_m}$, we have at each step only a single term of the quotient to be calculated. If, indeed, the polynomial coefficients P_m and Q_m change at each iteration, the first $m - 1$ terms of the quotient giving R_m are the same as the first $m - 1$ terms of the quotient giving R_{m-1}.

In the application made, the calculations are relatively long and, whereas good results are obtained on synthetic seismograms, present ignorance of the exact surface conditions means that on actual seismograms the results are rather misleading. In other respects, because errors are committed in the evaluation of certain reflection coefficients, arising particularly from the presence of noise of any kind, it is found that this often brings about discrepancy, which tends to increase with time, between results obtained and those anticipated. The demultiple operation should therefore only be carried out on traces previously refined to the utmost from noise.

Likewise, we have investigated whether approximate procedures not applying the rigorous condition would give analogous results.

We can, in fact, suppose that an elaborate deconvolution performed on the spectra might have the effect of diminishing the amplitudes of spectra expressing the presence of multiple reflections and, in that way also, attenuate or suppress them. The tests made have been carried out by operating only on the real part of the trace spectra. We take the mean of the spectrum which closely assumes the fluctuations of the absolute value of the real part. In this manner it is considered that the implied impulse is an autocorrelation since its spectrum is positive. Remaining in the frequency domain, the term-by-term division of the real part of the trace spectrum is made by the mean obtained. We then pass back into the time domain through the inverse Fourier transform.

In part *a* of Figure 58 an original trace is shown, in part *b* the result of the previous operation (deconvolution). A rigorous demultiple operation has been applied to the result of *b* and we observe, in part *c*, that the trace obtained is practically identical to the deconvolved trace. An alternative is introduced by the fact that the demultiple operation has brought about nothing new: the elaborate deconvolution either suppresses the multiples, or there are no multiples on the trace considered.

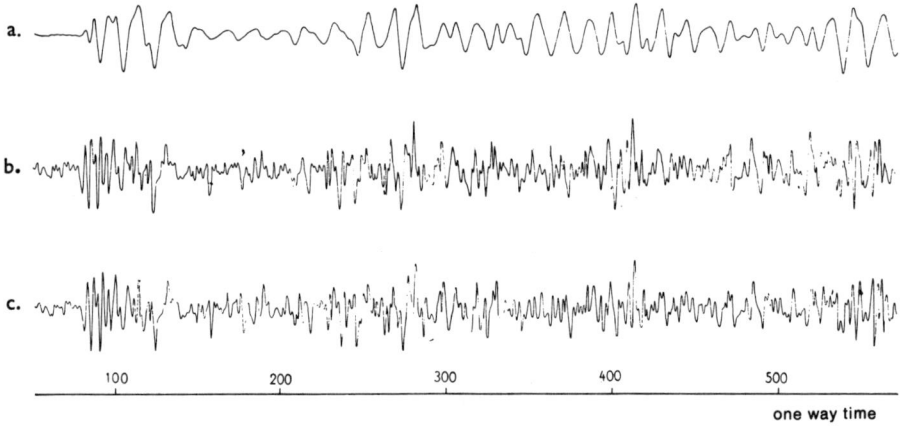

Figure 58.   An original seismic trace (a), together with the corresponding deconvolved (b) and filtered by demultiple process (c) traces.

We then applied this deconvolution procedure to a marine seismogram where the presence of multiples is obvious (Figure 59a). A filter of the Backus type gives the result shown in Figure 59b. The elaborate deconvolution procedure applied to the original record gives the result shown in Figure 59c. The similarity between Figures 59b and 59c can be observed.

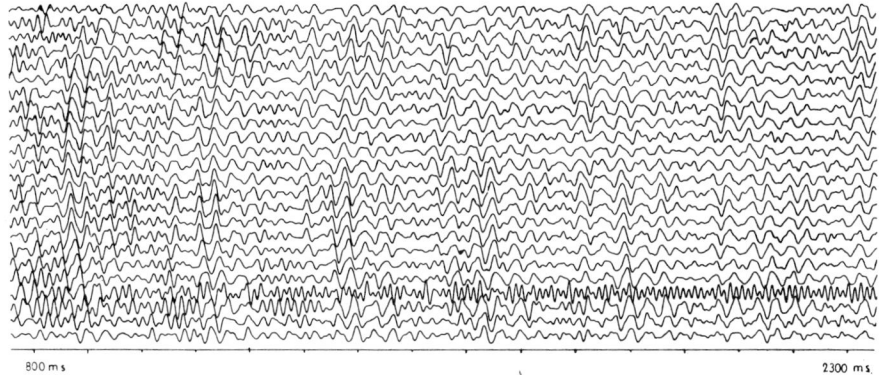

800 m s   2300 m s

Figure 59a.   Original marine record displaying obvious multiples.

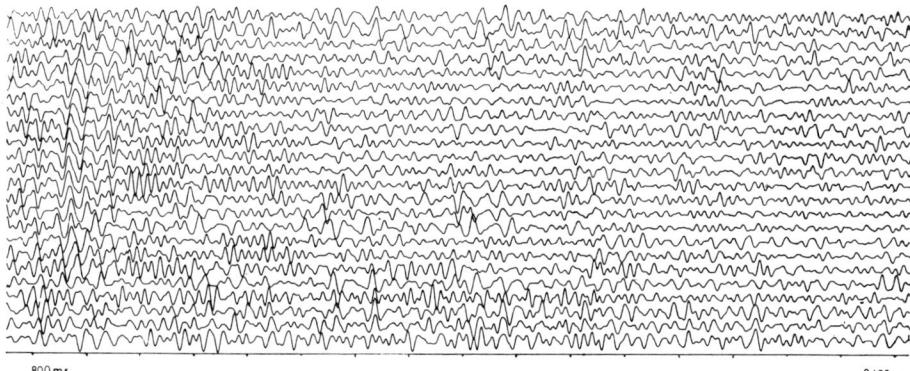

800 m s   2400 ms

Figure 59b.   Record obtained from (59a) by Backus type filtering.

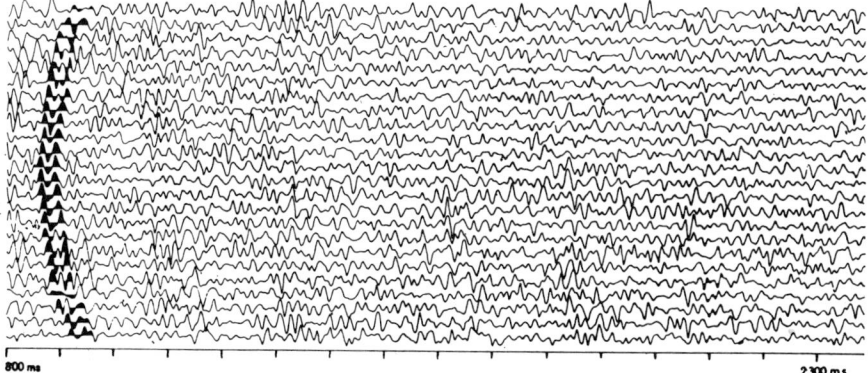

800 ms   2300 m s

Figure 59c.   Record obtained from (59a) by strong deconvolution.

It is clearly seen that this procedure removes only the multiples whose importance is such as to be expressed by spectral amplitudes higher than the mean amplitude in a small band of frequencies around the resonant frequency. The advantage
of the procedure is in not requiring knowledge of any filtering parameters; its
drawback is in being rather cumbersome and difficult to employ in routine work.

Theoretically, it seems very daring, even totally discrepant, to filter the real
part of the trace spectrum by the mean of the modulus of this real part. It should
be necessary to do the filtering through the mean absolute value of the trace spectrum. Actually, the two procedures give perceptibly the same result, which is
tantamount to saying that the phase does not have great importance. This already
stands out in the example given in the first chapter, where starting from only the
modulus of the trace spectrum, we obtain practically the trace itself. Likewise, we observe that if we filter a marine trace with a symmetrical zero phase
operator whose spectrum has a modulus equal to that of the Backus type operator, we obtain almost the same result. This shows that seismic traces display
peculiarities which have not yet been completely brought out.

Chapter 7

ANALOG FILTERING OF MARINE SEISMIC RECORDS

B. DAMOTTE

Record sections obtained from a marine seismic survey are sometimes rendered unusable by disturbances of periodic nature due to the water layer. The latter is, in fact, bounded by two interfaces with strong reflection coefficients, the air-water interface with coefficient -1, the water-substratum interface with a coefficient value that can reach 0.8.

The seismic energy produced in this layer or arriving there from below is successively reflected by the two interfaces with a time interval equal to twice the thickness of the water layer traversed at the velocity of sound in water, namely 1500 m/sec.

In deep water, $Z > 30$ m, this is expressed by multiples or reverberations whose separations in time represent the forward and return travel path in the water. This kind of section is shown in Figure 60a. The water depth is 45 m. To the left of SP 1706, the intensity of the reverberations completely obscures the reflecting horizons.

In shallow water, $Z \ll 30$ m, the reverberations being as close together as possible give rise to apparent sinusoidal phasing, a phenomenon often termed "singing". Figure 61a is an illustration of this. The water depth is less than 10 m and the dominant frequency occurs toward 40 hz.

First we will describe the method that we have applied in order to make these sections useful. Next we will set forth some results based on experience, limited to the analog filtering of reverberations due solely to the presence of a water layer above a sedimentary series capable of originating a complete train of multiple reflections. Figures 60b and 61b show the results obtained starting from Figures 60a and 61a.

In the case of reverberations (Figures 60a and 60b), filtering has had the effect, in addition to suppression of multiples, of permitting a more continuous and certain correlation of horizons at 1 sec and 1.8 sec; these were interrupted at SP 1705. Between SP 1699 and SP 1705, a region where shallower horizons appear, filtering has allowed us to distinguish the true horizons among the multiple horizons and to detail a convergence between the horizons at 1650 msec and 1800 msec at SP 1699. Futhermore, a constant character exists on the horizon at 1800 msec, which facilitates precise correlation on either side of zones of maximum interference where this horizon is weak (1706-1707, 1710-1711).

Figure 61b presents the result obtained in the case of singing from Figure 61a. Here again some horizons appear (1200 msec,[10] 2000 to 3000 msec) after filtering.

[10] Millisecond is rendered "ms" in the figures and tables in this chapter as in previous chapters.

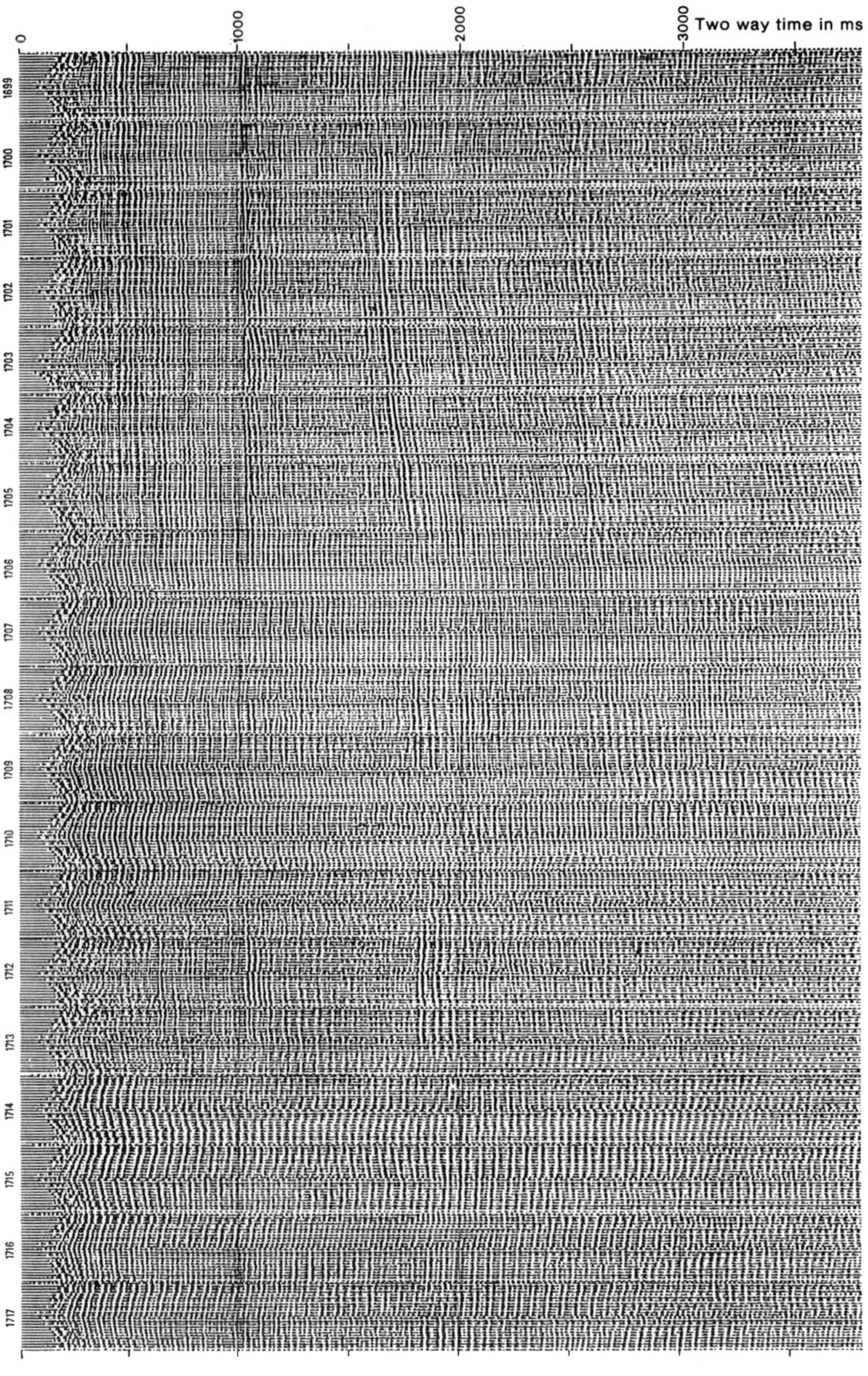

Figure 60A.   Marine record section showing ringing before filtering. Recording bandpass 18-75 hz.

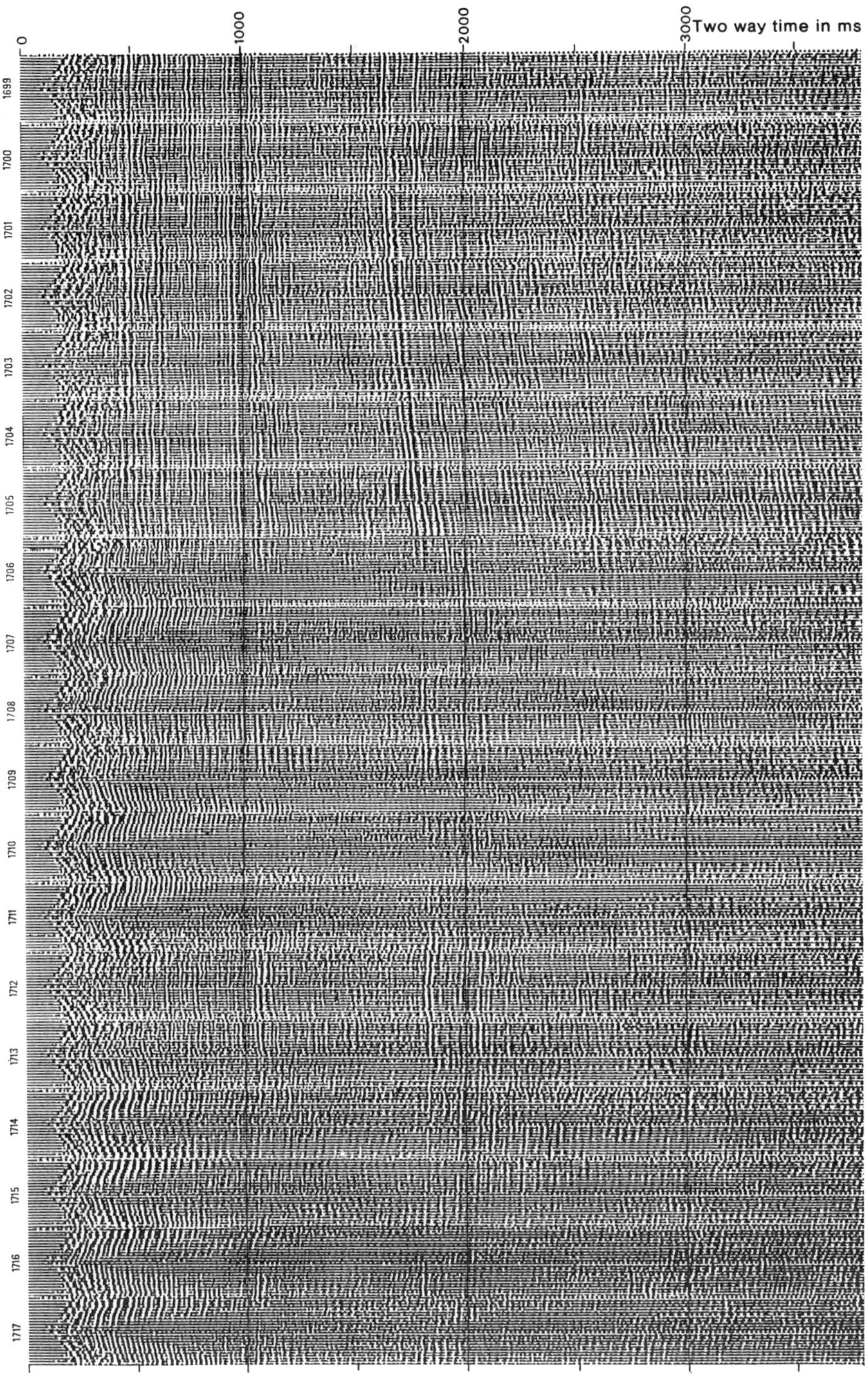

Two way time in ms

Figure 60B.   Record section from Figure 60A filtered to eliminate reverberations.

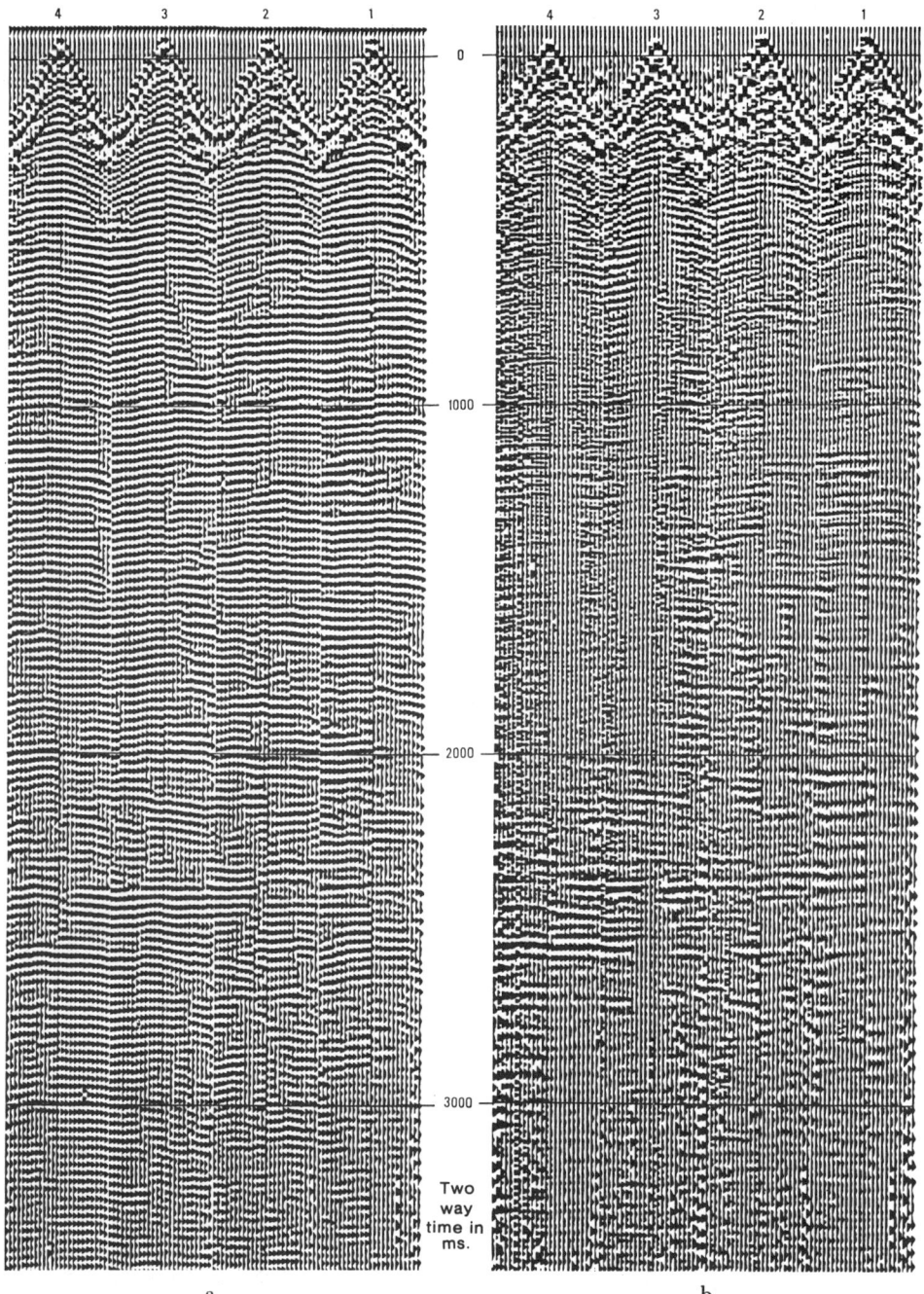

a

b

Figure 61.   a) Marine record section showing ringing before filtering. Recording bandpass 18-75 hz. b) Record section from Figure 61A filtered to eliminate ringing.

FILM STUDY. AUTOCORRELATIONS

In order to arrive at these results, a prior study of the original films is necessary.

There is often the tendency to filter marine playbacks with a low-pass filter which suppresses all frequencies above 30 or 40 hz, and which has the effect in shallow water mainly of accentuating the ringing and of making such sections even more illegible than if they had been filtered with the broadest band possible (O-92 or 120 hz, for example).

As an example, Figure 62 shows the same original section filtered with the bands 0-40 hz and 18-92 hz. The section with 0-40 hz filter displays cyclic behavior. The dominant frequency is located between 30 and 33 hz. Three zones of energy can be distinguished from 400 to 900 msec, then from 1300 to 1600 msec and finally around 1900 msec. With the exception of this last level, it is difficult to define a correlation from the viewpoint of phasing choice as well as its continuity; only a south dip is clear.

The simple action of "opening" the electrical filter clears up the section in a passable fashion. In the first energy zone we find a convergence; a high-frequency phasing at 600 msec at SP 10 comes to rest against a subhorizontal arrival at 800 msec at SP 11. In the second energy zone, an arrival dipping toward the south begins at 1400 msec (SP 10) and can be traced up to SP 12. Finally, the horizon located around 1900 shows up better and even displays character which gives it a less cyclic behavior.

Another example is shown in Figure 70 where the appearance of character and enhancement of reverberations can be seen by the simple action of passing from an 18–48 hz filter on the monitor to a 21–66 hz filter on the section.

An uncorrected version, filtered solely because of high-frequency instrument noise, consequently enables us to refer the problem to its exact category. We examine this material to localize, in place and time, portions of the section which must be analyzed in detail by means of autocorrelation.

The latter is noteworthy because it clearly indicates all periodic phenomena which are more or less subdued on normal films and thus enables us to study at leisure the nature and period of the extraneous events. Autocorrelations are rapidly made with the aid of the IFP optical correlator.

Two specimens of the trace to be studied are copied over a transparent base. The window of interest to us, taken on one of the samples, is fixed in a beam of collimated light in normal incidence. The other specimen is steadily played in superposition over the first. The variation of light as a function of time resulting from this operation is measured by a photoelectric cell, the output current of which activates a galvanometer. Figure 63 shows the films and corresponding autocorrelations. From top to bottom:

1. *Ringing*

The phenomenon is already quite evident on the film, but its qualitative and quantitative study is more precise on the autocorrelation which displays an inter-

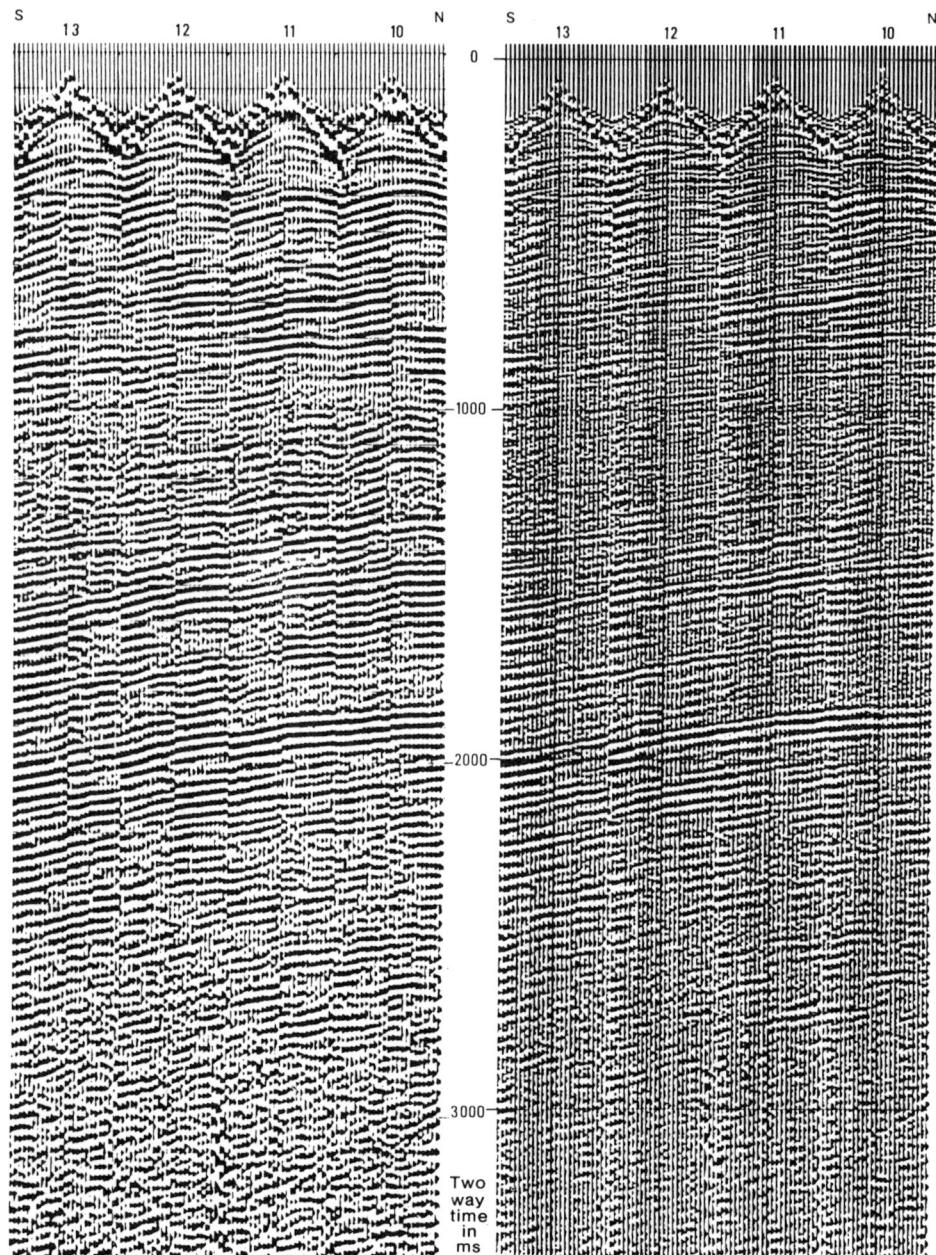

Figure 62.   Effect of low pass filtering 0-40 hz (left) and bandpass filtering 18-92 hz (right) of the same original record.

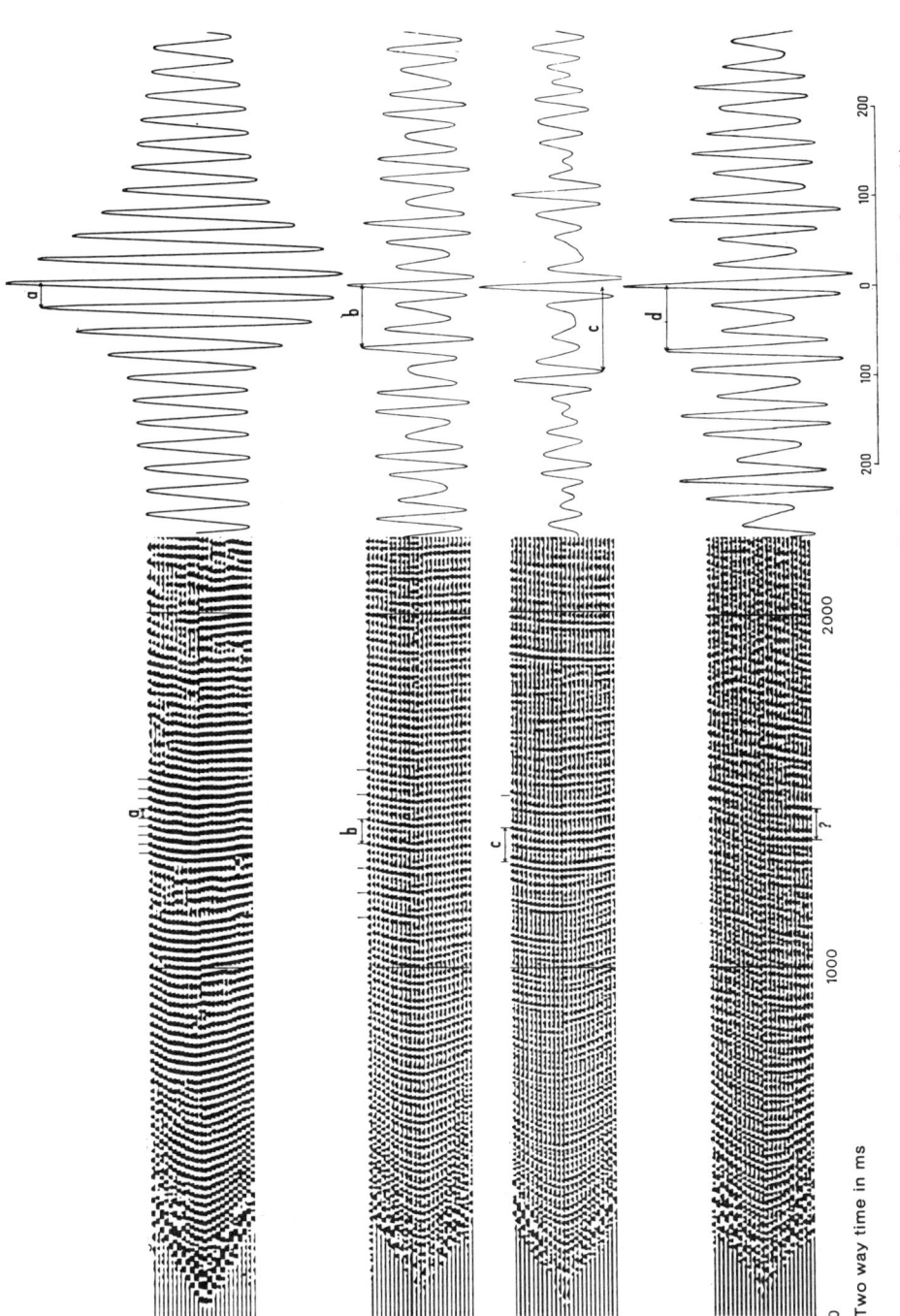

Figure 63. Examples of autocorrelation of marine records showing ringing in water 8m deep (a); and reverberations in water 45m deep (b), 65m deep (c), and 57m deep (d).

ference frequency of 40 hz. The autocorrelation was performed on trace 13 between 1 and 2 sec.

2. Reverberations

Here again the phenomenon is evident on the film, but autocorrelation defines it more clearly. We can thus see that at this shotpoint the dominant periodicity is a reverberation system in which the arrivals spaced 69 msec apart all have the same polarity. The periodicity corresponding to the water depth (45 m), i.e., 58 msec with alternating polarity of multiples, attenuates very quickly. The autocorrelation was performed on trace 9 from 1200 to 2200 msec.

3.

We are concerned here with a film whose repetitive character is less clear. Autocorrelation shows a reverberation with alternating sign at 92 msec, corresponding to the water depth, and also important, a reverberation of constant polarity at 102 msec (autocorrelation of trace 13 from 1200 to 2200 msec).

4.

Study of the disturbance on this film is made difficult by numerous arrivals and by noise. The autocorrelation here more than anywhere else facilitates a good identification and an accurate measure of the interference period, namely: reverberations of similar sign corresponding to the water depth (57 m) i.e., 76 msec and very subdued repetitions of alternate sign at 85 msec.

The autocorrelation was carried out on trace 9 from 1200 to 2200 msec.

REMARKS ON THE TRAVEL PATH OF SEISMIC WAVES

Detailed study of these autocorrelations, previously assembled in sections comparable to seismic sections, enable us to identify the disturbance and to calculate the parameters of the filter operator necessary for its elimination. In the case of reverberations, this operator is generally based on the period of the interference which is easily measured on the autocorrelations. If operators of this type seem simple in both their principle and application, it does not appear less so that the result of their routine use is often misleading. The quality of this result is a function of the constancy of the period in time and space, and of the accuracy with which this period is known, if the same operator is applied to several traces, even to several films.

Before describing the method of filtering that we have employed to obtain the results of Figures 60b and 61b, we are going to study in the examples the factors influencing the variation in period, then the accuracy required in order for the filtering to be effective. Finally, we shall see in what sequence we have to develop the operations of filtering and dynamic correction.

For this we are led to make assumptions on the wave travel paths in the water and the substratum. These assumptions are illustrated at SP 1720 (Figure 64) and are verified on all records that we have filtered.

Let us study the time variation of the Δt^{11} at this shotpoint. Note first of all

[11] Δt is understood to be the difference between the arrival time of the same reflection on the inner and outer traces, i.e., between traces 12 and 24.

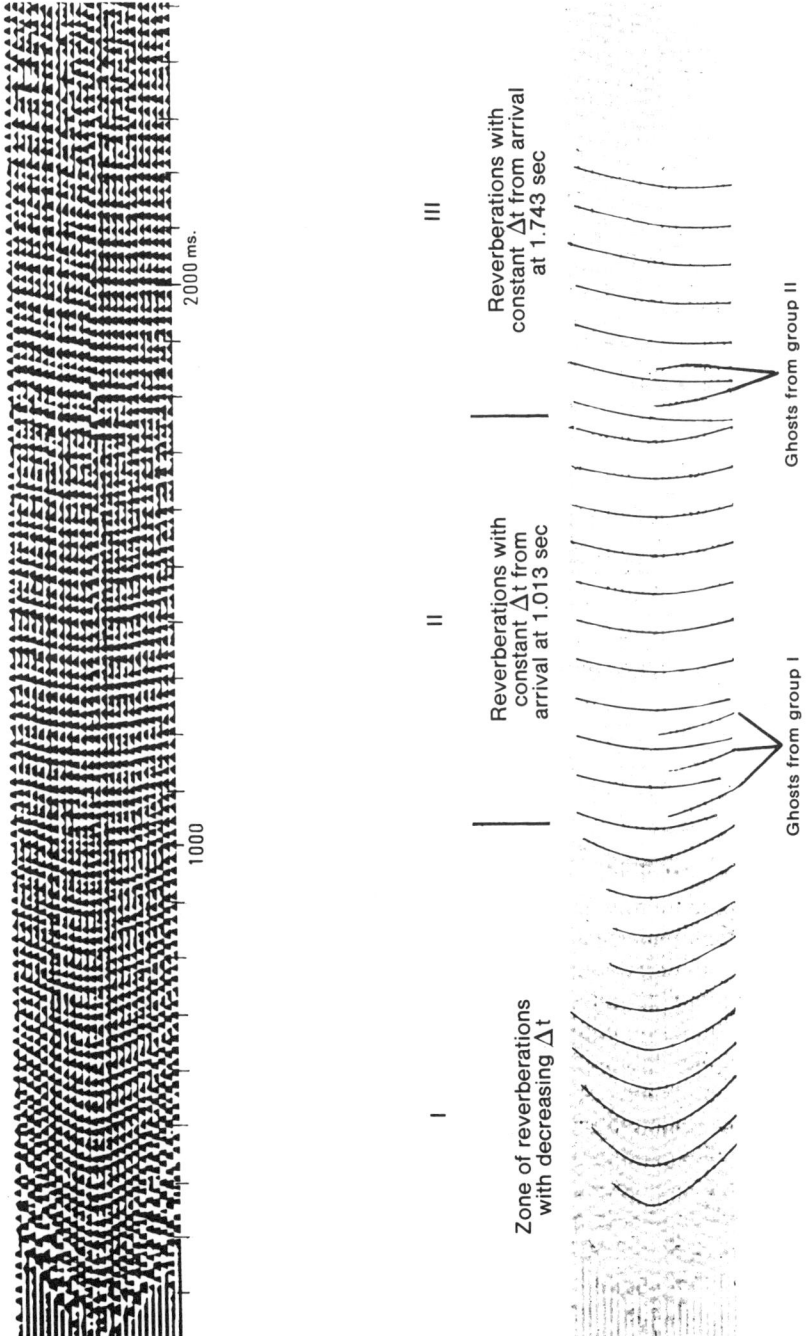

Figure 64.   Move out study on original record from shot 1720.

that on this film the reverberations do not change sign.

This film can be divided into three segments:

1) From 0 to 1013 msec:

Reverberations spaced at 69 msec on the inner traces present steadily diminishing Δt as a function of time. Each reverberation corresponds to a direct propagation and additional return of the seismic wave between the two surfaces of discontinuity. Increase in the number of direct and return travel paths is equivalent to greater and greater distances covered and brings about a decrease in Δt corresponding to constant V of 1500 m/sec (Figure 65).

2) From 1013 to 1743 msec:

A wave reflected from a marker occurs at 1013 msec. This wave has traversed the water layer, then the geologic section going and returning, i.e., a medium where the average velocity can reach 2500 m/sec.

The Δt for $V = 2500$ m/sec at 1013 msec is clearly less than the Δt for $V = 1500$ m/sec at that time. An abrupt break results from this on the curve $\Delta t = f(t)$. The change of curvature is very well observed on the film (Figure 64). Next a complete series of reverberations enter until 1743 msec. The spacing of these events is apparently the same on the inner traces as well as on the outer ones. In other words, the Δt no longer seems to vary as a function of time and would appear to keep the value that it assumed at 1013 msec with the arrival of the reflected wave.

In effect, the seismic wave coming from deep horizons arrives at each trace following an almost vertical travel path with a Δt depending on the average velocity and time. It then continues to execute direct and return traverses in the water layer at the vertical of each trace, just as if each trace had become a source of energy, successively displaced by the Δt adjusted to the time 1013 msec.

The curve $\Delta t = f(t)$ therefore appears horizontal up to 1743 msec.

3) Starting from 1743 msec:

The process is renewed by means of another energy arrival, this time with dip. There is an abrupt decrease of Δt on the curve and on the film, and the former again remains constant with its new value; hence a new horizontal stage on the curve.

It is interesting to note that when a sequence of reverberations with constant Δt overlie another sequence, we can still find ghosts of the previous group in the form of interference in the new series.

In summary, at the instant of a marine shot, part of the emitted energy finds itself channelled in the water layer and generates periodic arrivals. The period on the closest traces to the shot corresponds to twice the thickness of the water layer traversed with a velocity of 1500 m/sec. On the film ensemble these arrivals have progressively weaker curvature corresponding to propagation in a homogeneous medium with $V = 1500$ m/sec.

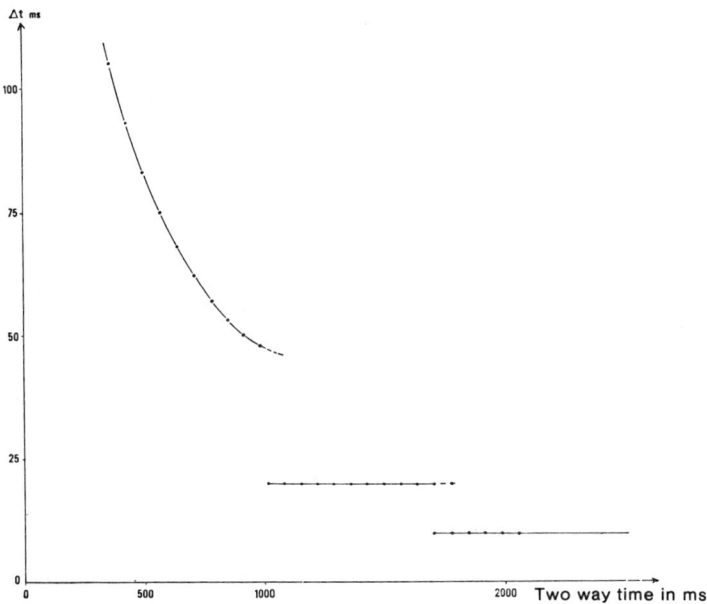

Figure 65. Move out study from shot 1720.

The remainder of the energy which has penetrated the geologic section, and is reflected from a marker, undergoes reflections in the water layer on its return, generating a new series of periodic arrivals. Among this phasing, only the first is valid and the subsequent ones appear parallel to it.

This parallelism between multiple arrivals, or this constancy of Δt in time, when it is a question of reverberations because of the water layer, is only apparent. At SP 1720, it is due only to the coincidence of the following facts: a short spread (420 m), water depth fairly important (45 m), arrival time of the original horizon greater than 1 sec. Actually these Δt's vary slowly and the difference between the Δt of the true horizon and that of its first multiple, a difference that is denoted by $\delta \Delta t$, is, under the conditions mentioned above, less than a millisecond, hence its apparent constancy.

A theoretical study, treated in detail in the Appendix, shows that $\delta \Delta t$ is a function of four parameters: the water depth Z, the length of the geophone spread X, the arrival time T_1 of the horizon which originates a train of reverberations, and finally the average velocity V_m which characterizes the strata between water bottom and the horizon measured at the time T_1. The difference $\delta \Delta t$ increases with Z and X and decreases as T and V_m become larger. In the appendix to this text, values of $\delta \Delta t$ are given for different ranges of the parameters Z, X, T_1, and V_m.

It will be shown that, in order for filtering to be valid, it is necessary for the reverberation period to be known with an error less than one msec. Being given this condition, the effectiveness of the filtering decreases as $\delta\Delta t$ becomes larger. In practice, when $\delta\Delta t$ is less than one msec, we can consider the multiples from the horizon to be parallel.

This occurs for short spreads (less than 400 m) and ordinary values of Z, V_m, and T_1. All the records shown here fall into this category.

Indeed, as soon as the first horizon has taken the place of the water reverberations, the time difference between each multiple arrival rapidly becomes less than 1 msec and is no longer a function of the shot to detector separation, but depends essentially on the seismic sea bottom depth for the trace considered. The time difference between multiples is therefore the same for a given trace, but can vary from one trace to another if the seismic sea bottom depth changes.

On the other hand, the dip of a subsurface horizon (provided that it is not too large) does not alter the time difference between multiples to the extent that the seismic sea bottom is flat.

It is therefore essential not to destroy the natural regularity of differences between multiples by dynamic corrections, which are always approximate. Moreover, it is much simpler and more rigorous to define a velocity function from Δt statistics on a record free of water derived multiples.

It is evident that multiple arrivals originating from travel paths conducted solely in the water layer and conforming to the law of curvature decrease will be only partially filtered. The filter will take effect only insofar as the time difference is equal to or almost equal to the time difference between multiples generated by arrivals from deep horizons; i.e., on the inner traces, then progressively toward the side traces as a function of time.

In contrast, in the case of long spreads (1200 m and even 600 m) the fact that $\delta\Delta t$ is greater than 1 msec makes it difficult to apply any means of filtering which calls on the periodicity of multiples.

We can, as in the case of short spreads, filter before applying dynamic corrections, but given that Δt varies in time and space it is necessary to change the filter with each trace and to make it vary in time and space in appropriate fashion with each film.

An alternate method consists of correcting dynamically before filtering in order to bring the different horizons and their multiples parallel. It is therefore necessary to study each film in advance so as to know which procedure is suitable for it; this requires that the films have a good signal-to-noise ratio. Thus for SP 1720, previously studied, it would be necessary to correct with a constant 1500 m/sec up to 1013 msec, then with a relation correcting the same Δt from 1013 to 1743 msec and lastly with a third relation after 1743 msec.

Finally, if the correcting procedure is successful, the filtering will be easy and if weaker horizons appear previously masked by multiples from stronger horizons, they will come out overcorrected and will then require additional dynamic correction.

This method is not practical. In the case of long spreads it seems that we are led to conduct filtering only on well-defined horizons, and this to the detriment of the others.

Comment concerning Δt statistics performed on marine shooting

The phenomenon presented here for three beds at SP 1720 may be an explanation for anomalies found during analysis of the Δt's of marine shooting. Figures obtained on this shooting give neither the water velocity nor the relation found in a neighbouring well, but something intermediate in the form of an initial grouping of points around $V = 1800$ to 1900 m/sec, and then points progressively more dispersed around an average curve $V = 2200$ to 2400 m/sec.

Let us expand the present case to an infinity of layers behaving in the same manner; we will obtain a film where the true arrivals interfere with reverberations of slightly variable Δt originating from previous arrivals. The result of a Δt analysis will therefore be incoherent.

FILTERING WITH A TWO-SAMPLE OPERATOR

The simplest of operators has been applied to films where the periodic phenomenon is very pronounced. It consists of adding to each trace the same trace suitably adjusted in sign and delayed by an amount equal to the period of the phenomenon.

The operation is developed on the IFP delay-line filter. We introduce the original trace on the first drum, adjusted to minimum delay. The same trace is played on a second drum, delayed with respect to the first by the desired amount. The appropriate sign and weighting are impressed on these traces at the time of demodulation.

In order to estimate the accuracy necessary for measuring the period, we have first of all filtered the same shot with the delay measured on the films or on the autocorrelations, then with delays of $\pm 1, \pm 2, \pm 3$ msec around the value found.

These results are brought together in Figures 66, 67, 68, and 69.

For each of these figures, a single delay value suffices; these are:
— Figure 66: 69 msec (SP 1707)
— Figure 67: 69.5 msec (SP 1720)
— Figure 68: 68 msec (SP 1731)
— Figure 69: 92 msec (SP 5370)

On Figures 66, 67, and 68, where the reverberations are very pronounced, it can be seen that a 1-msec displacement on either side of the exact value is sufficient to cause the multiples to reappear. The margin of error can therefore be estimated at ± 0.5 msec in order for an operator with a sampling interval equal to the period, to be effective.

It is naturally not a question of repeating this sequential filtering for each shot. On the other hand, considering the extreme simplicity of the operation, it is very easy to adjust the delay on the delay-line filter itself.

For this we simultaneously display the trace to be filtered and the output trace of the delay-line filter on a multichannel oscilloscope screen. It suffices to vary slowly the delay introduced on the second ring of the delay-filter and to observe the effect produced on the screen. It takes about a minute to find the exact value of the delay.

For marine profiles shot over sea bottoms, with perceptibly constant depth measured by sounding, we can consider that the sea bottom is horizontal for each individual shot. Also, the delay may be assumed equal for the 24 traces. This delay is determined visually on the oscilloscope from one trace. The position of this trace does not matter if the reverberations result from a seismic reflection. On the other hand, if they are caused by the direct wave arrivals traveling only in the water, it is necessary to take the trace closest to the shot.

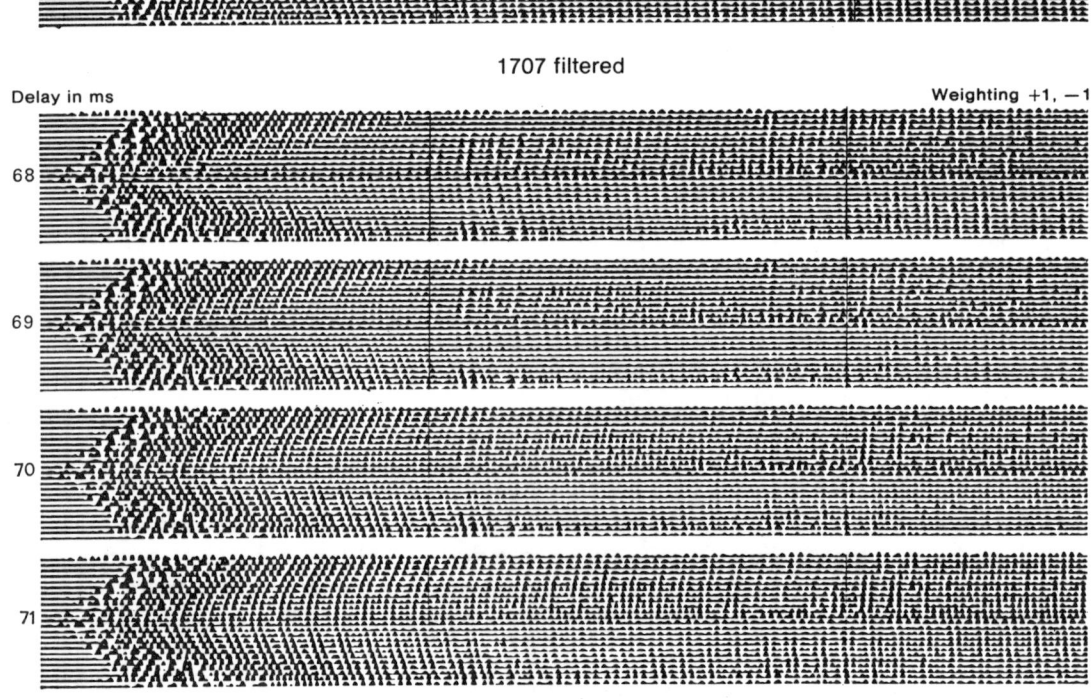

Figure 66. Study of delays on shot 1707, reverberations with the same sign. Recording bandpass 18-75 hz.

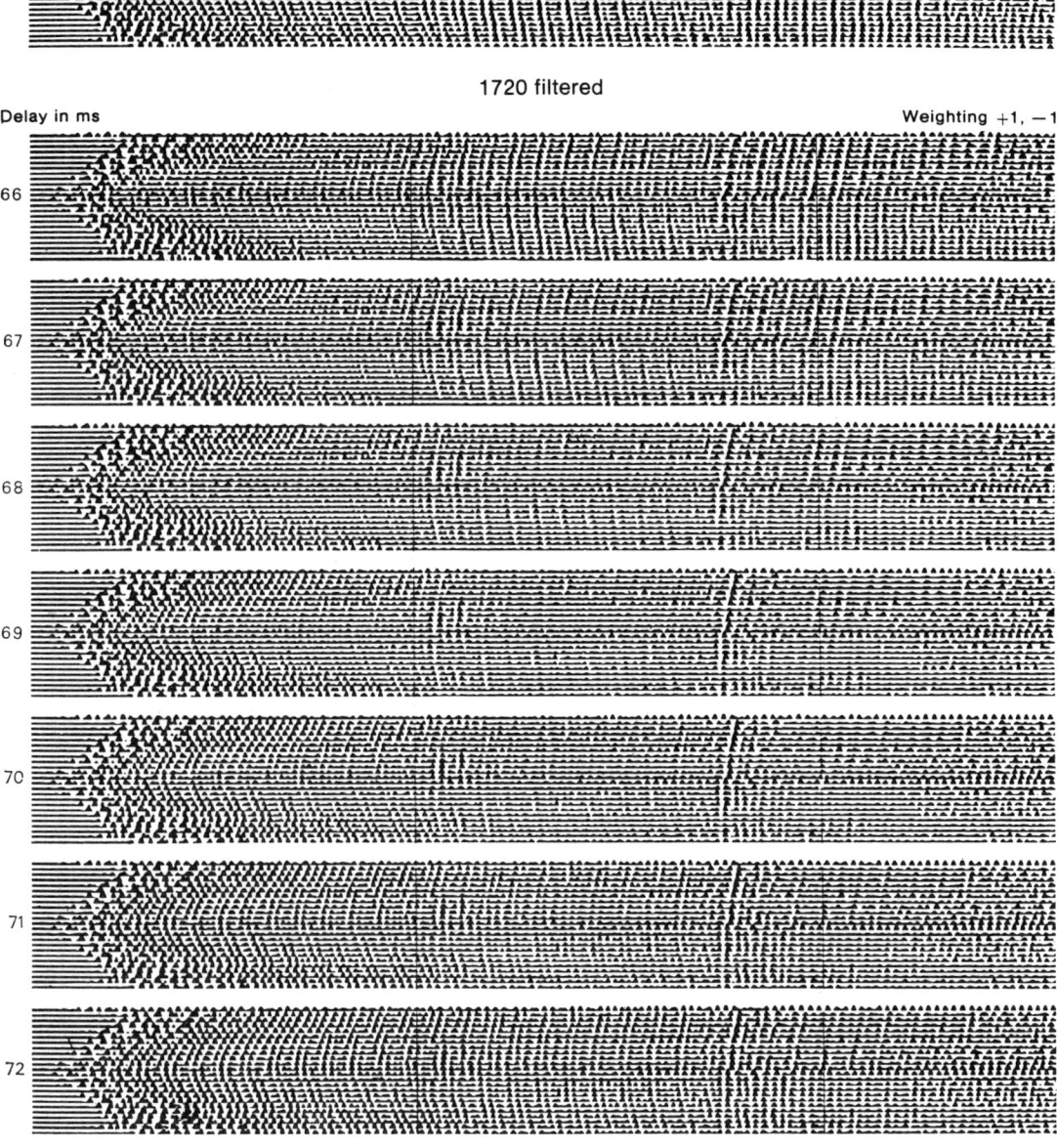

Figure 67. Study of delays on shot 1720. Recording bandpass 18-75 hz. Reverberations of same sign.

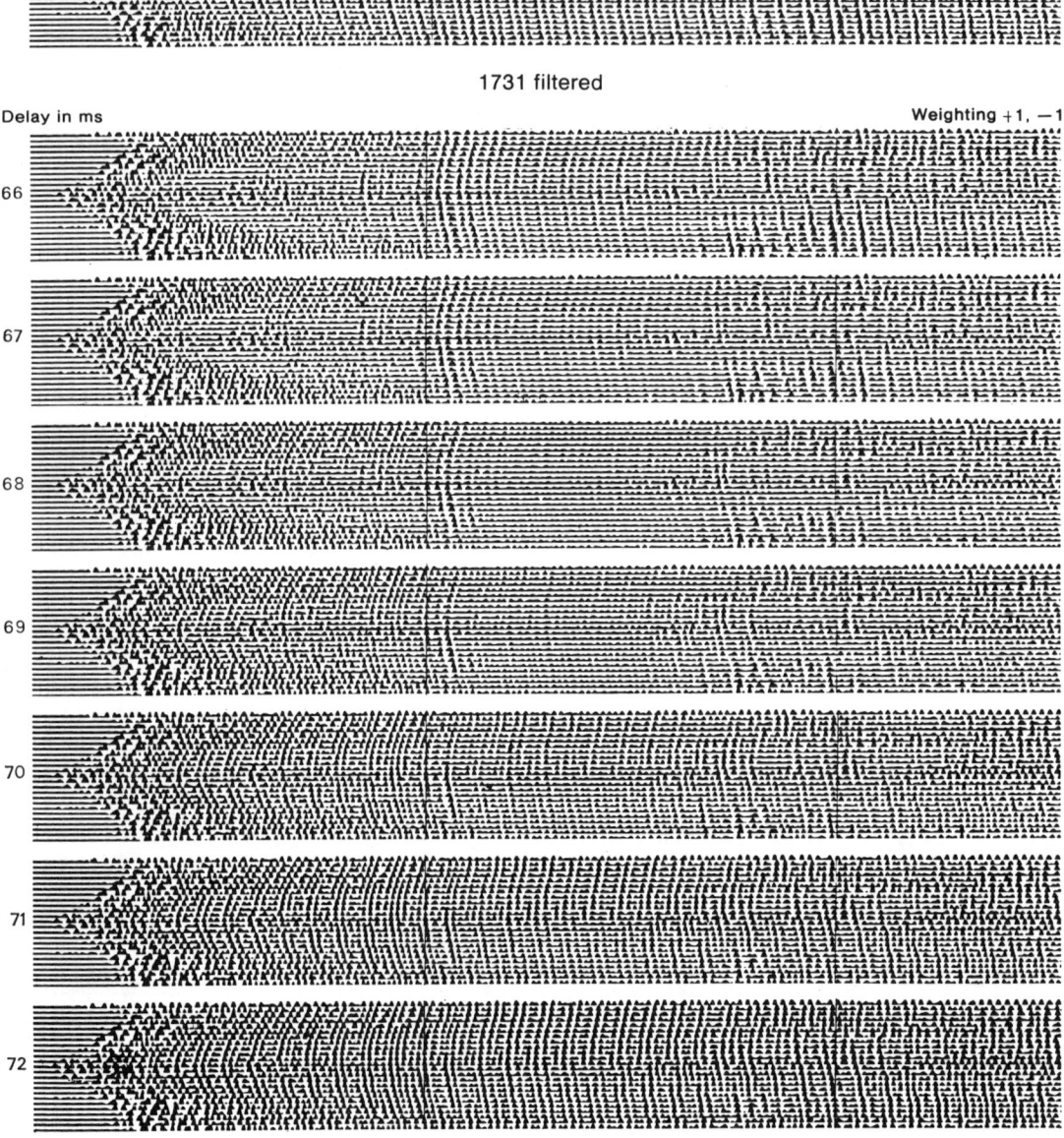

Figure 68. Study of delays on shot 1721. Recording bandpass 18-75 hz. Reverberations of same sign.

5370 original

5370 filtered

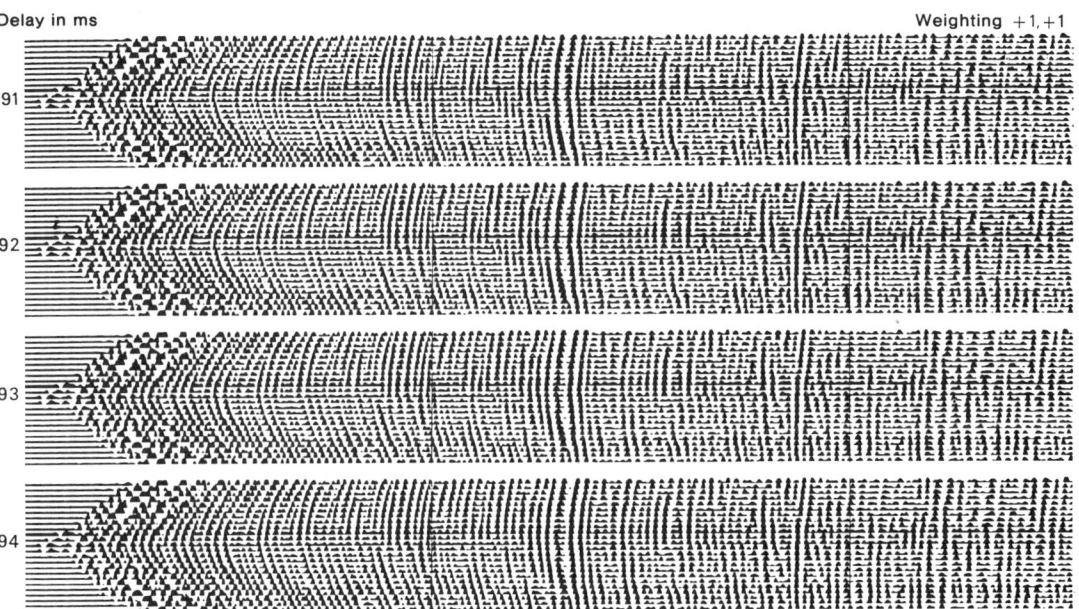

Delay in ms   Weighting $+1, +1$

91

92

93

94

Figure 69.   Study of delays on shot 5370, reverberations with alternating sign.
Recording bandpass 18-92 hz.

In areas where the submarine topography is variable, this type of filtering should be carried out by trace groups, or even trace by trace.

The advantage of this procedure, adaptable in the limit to each trace, is in providing constant results; i.e., to retain the horizon character on a large number of films, whatever the fluctuation in interference due to the water layer may be. This is in contrast to operators tailored to a group of shots with related parameters, then suddenly changed to accommodate another group of shots. The results of filtering are uneven in the same group, a particular shot not necessarily accommodating the same operator as the next one, there is furthermore the risk of a sharp break in horizon character and continuity, on changing the operator.

EXPERIMENTAL RESULTS

We have applied this straightforward method to several hundreds of marine shots from different areas. Previous study of these shots and study of the filtered versions enable us to present some experimental results.

Operator calculation from water depths

To convert these depths to time, it is necessary to utilize the water velocity. Generally, 1500 m/sec is taken, but this velocity can vary ± 50 m/sec according to the temperature and salinity of the water (i.e., an error of ±1.6 msec per period for a 40 m bottom).

A stastical study has been made covering a profile of 81 shots taken over sea floors with depths from 30 to 35 m and smooth topography. This study consists of comparing the time, calculated from the depth sounder with V_{water} = 1500 m/sec, to values of the period found by the delay-line filter and regarded as valid.

Table 1. Comparison of depth sounder to delay-line filter

Error (in ms) /Delay-line filter values	Number of Shots	(%)
0 to ± 0.5	33	40.75
± 0.6 to ± 1	18	22.2
± 1.1 to ± 1.5	9	11.1
± 1.6 to ± 2	11	13.6
± 2.1 to ± 2.5	7	8.65
± 2.6 to ± 3	2	2.45
± 3.1 and greater	1	1.25
	81	100.00

If those with an error less than or equal to ± 0.5 ms only are considered to be well filtered, it is seen that 40.75 percent of the shots would be correctly filtered by this method of calculation, which would present a profile of uneven quality, about one film in two still being affected by interference.

An identical study covering 52 consecutive shots at constant bottom sounded depths (54 m), gives the following results:

Table 2. Same comparison for constant bottom sounded depth.

Error (in ms) /Delay line filter values	Number of shots	(%)
0 to ± 0.5	12	23.0
± 0.6 to ± 1	15	29.0
± 1.1 to ± 1.5	4	7.5
± 1.6 to ± 2	10	19.5
± 2.1 to ± 2.5	4	7.5
± 2.5 to ± 3	3	6.0
± 3.1 to ± 3.5	4	7.5
	52	100.0

These 23 percent good results show how inaccurate the sounder indications are and that, if we wish to avoid oscilloscope adjustment, it is necessary to have a device giving the value of the period within a fraction of a millisecond.

The velocity in water has been taken equal to 1500 m/sec. The same statistical study has been made with $V = 1550$ m/sec, the percentages are slightly less for the error band 0 to ± 0.5 msec; i.e., 31 percent instead of 40.75 percent and 15.25 percent instead of 23 percent.

Difficulties in filtering as a function of signal-to-noise ratio in marine work

Let us denote:

  S: the seismic arrivals
  M: multiples
  B: incoherent noise

Marine records can be classified into four categories, according to the value of the ratio $(S + M)/B$. The progress and ease of filtering as well as its effectiveness depends on the value of this ratio:

$$\frac{S + M}{B} \gg 1 \quad \text{with} \quad M \gg S. \tag{1}$$

These are records where reverberations or singing exist in an almost pure state (Figure 60a, SP 1707; Figure 61a, SP 1). It is probable that the reflection coefficient of the marine bottom should be strong, generating first of all a long train of reverberations in the water layer. It is necessary then for the signal to be very powerful in order to substitute its own train of interference on previous data. A perceptibly constant level of interference is observed on these records that can be attributed to the combined action of a strong water reflection coefficient and AGC. The apparent reflection coefficient in this case is close to 1.

Autocorrelation is not mandatory for such records, the analysis of the phenomenon is capable of being made on the records. Furthermore, the oscilloscope adjustment is very simple,

$$\frac{S + M}{B} \gg 1 \quad \text{with} \quad M \neq S \qquad \begin{array}{l}\text{(Figure 69, original; Figure 60a,}\\ \text{shotpoints 1699 to 1705).} \end{array} \tag{2}$$

The reflection coefficient of the sea bottom should be weaker and only generate a suite of rapidly attenuated multiples. A larger number of seismic arrivals can be seen and are interspersed with multiples of previous arrivals. Autocorrelation is required every four or five shotpoints in order to measure the period. It is then necessary to register some multiples on the record, so that they can be utilized in adjusting the filter:

$$\frac{S + M}{B} \neq 1 \qquad \text{(Figure 63, bottom shot).} \tag{3}$$

Multiples are visible, but only by continuity on several traces. They can not be identified with certainty. Autocorrelation is necessary at each shotpoint in order to interpret the phenomenon clearly:

$$\frac{S+M}{B} < 1. \tag{4}$$

In this case, seismic events and multiples are buried in random noise. Nothing coherent stands out. Autocorrelation at each shotpoint is mandatory here; autocorrelations at different time sectors or time gates and on several traces are sometimes necessary.

Difficulties in filtering keep mounting when the ratio $(S+M)/B$ decreases, the interference phenomenon being less and less discernible. In other words, when the random noise increases we come to the conclusion that the more a record displays reverberations or pure singing, the simpler it is to filter.

Autocorrelation studies

The periodic phenomenon caused by the water layer is interpreted most of the time by assuming an air-water interface reflection coefficient of -1, and a reflection coefficient from $+.1$ to $+.4$, and even more, for the water-sediment interface (for downgoing waves). For water depth greater than 30 m, this approach gives multiples with alternating polarity because of the different sign of the two reflection coefficients.

Certainly, examples conforming to this theory occur (Figure 71, top record, and Figure 70), but from autocorrelation study of numerous marine shots it is apparent that if this phenomenon exists, it is far from being simple.

Furthermore, it is not necessarily the water bottom (if such can still be defined in a muddy bottom) that returns the waves to the surface, but may be a shallower horizon. The observed period therefore has nothing more to do with the two-way traveltime pattern in the water.

Figure 70 shows an example of this. The three records displayed were shot at the same bottom depth, 79.25 m, which should give horizon repetitions spaced 106 msec apart with $V = 1500$ m/sec. But record 3105 and its autocorrelation, display a repetitive phenomenon with alternating sign every 137 msec, and the assemblage of three shots shows that shot 3104 has more frequent repetitions (135 msec), and for shot 3106 they are spaced further apart (140 msec). The result of this is a slope increasing with time (on the three shot assemblage, the sign change of the repetitions, and the slope are marked).

Apparently the reflector generating the disturbances is slightly inclined and is located in a lithologic series below the water bottom, with a depth that cannot be precisely defined, since the average velocity between water bottom and this reflector is not known.

The detailed study of autocorrelations reveals for the majority of deep water shots that we have had to filter two periodic phenomena:

1) a system of reverberations of alternating sign with period T;
2) a system of reverberations of the same sign with period T'.

The smaller of the two periods corresponds to the water depth; its other period is about 10 msec larger.

In the case where the reflection coefficient at the bottom of the water is positive (the most frequent case), T corresponds to the depth of the water and we have $T' > T$. Conversely, if this coefficient is negative (much rarer case) T' corresponds to the water depth and we have $T > T'$.

Such a phenomenon, which appears only on the autocorrelation and which partly influences the manner of filtering, must probably be associated with the shape and spectrum of the impulse traveling in the water.

An example of a shot taken above a marine bottom with negative reflection coefficient is displayed in Figure 63, lower record. Its autocorrelation shows a preponderance of reverberations of the same sign, the period (76 msec) of which corresponds to the water depth (57 m), and some reverberations of alternating sign, very rapidly attenuated, having a period of 85 msec.

Inasmuch as the reflection coefficient at the sea bottom is most often positive, the detailed study of these two periodic phenomena is made on records shot above such sea bottoms. Figure 71 is an illustration of this.

A reverberation of alternating sign corresponding to the water depth and a reberberation of the constant sign with a period 10 msec greater are found for these three shots coming from very different localities:

SP   334:   sounder depth $Z = 54$m, $+ - - = 74$ msec $+ + = 85$ msec

SP 1707:   sounder depth $Z = 45$m, $+ - - = 58$ msec $+ + = 69$ msec

SP 5370:   sounder depth $Z = 65$m, $+ - = 92$ msec $+ + = 102$ msec

It is probable that these two phenomena evolve in relation to each other as a function of the marine bottom reflection coefficient value.

1) The reverberations of alternating sign dominate. This is the common case. The simplest filtering, as we have seen, consists of summing each trace with the same trace displaced by the period (Figure 71, 334).

2) Reverberations of constant sign are dominant. The same filtering is applied, but with sign inversion of the delayed trace (Figure 71, 1707).

3) The two phenomena are each as important as the other (Figure 71, 5370), in this case they can be filtered by either means, that is by applying the operator:

$$f_1(t) = g(t) + g(t - T),$$

or:

$$f_2(t) = g(t) - g(t - T'),$$

$g(t)$ being the original trace.

Figures 66 and 72 present a study of the delay in the two cases for shotpoint 1707. Both phenomena occur (Figure 71, autocorrelation), but the reverberations of constant sign are more important and the best filtering is that of Figure 66.

Figures 69 and 73 present a study of the delay in the two cases for SP 5370. Here both phenomena are equally important (see Figure 71, autocorrelation) and both kinds of filtering appear equally good. Nevertheless, for this last example, the best filtering consists of combining both kinds of operators, i.e.:

$$f(t) = g(t) + 1/2\ g(t - T) - 1/2\ g(t - T').$$

Original records

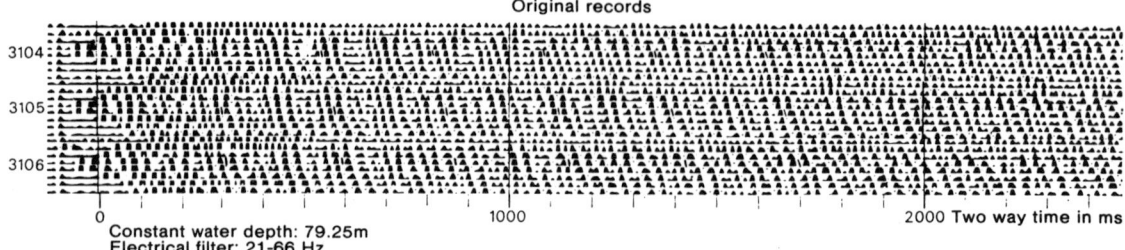

Constant water depth: 79.25m
Electrical filter: 21-66 Hz

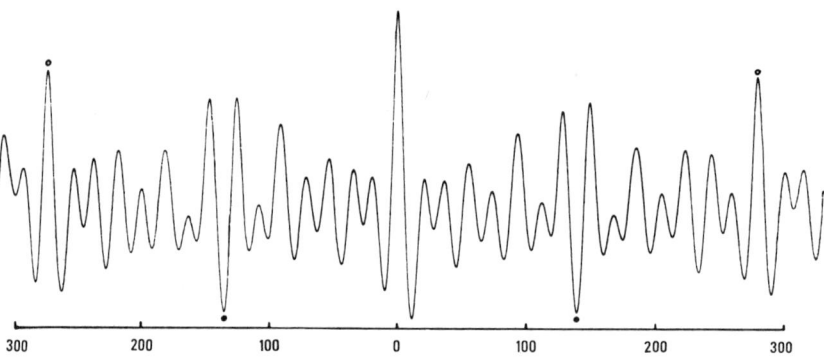

Autocorrelation of 3105

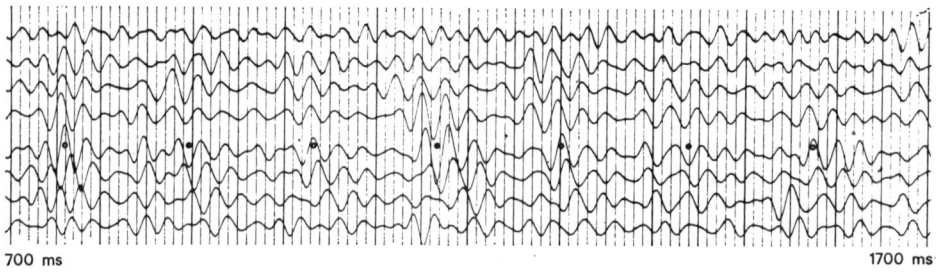

Portion autocorrelated from 3105

(Partial reproduction of monitor 18-48 Hz electrical filter)

Figure 70.   Example of reverberations independent of the water depth.

Sp 334

Sp 1707

Sp 5370

1000   2000

Two way time in ms

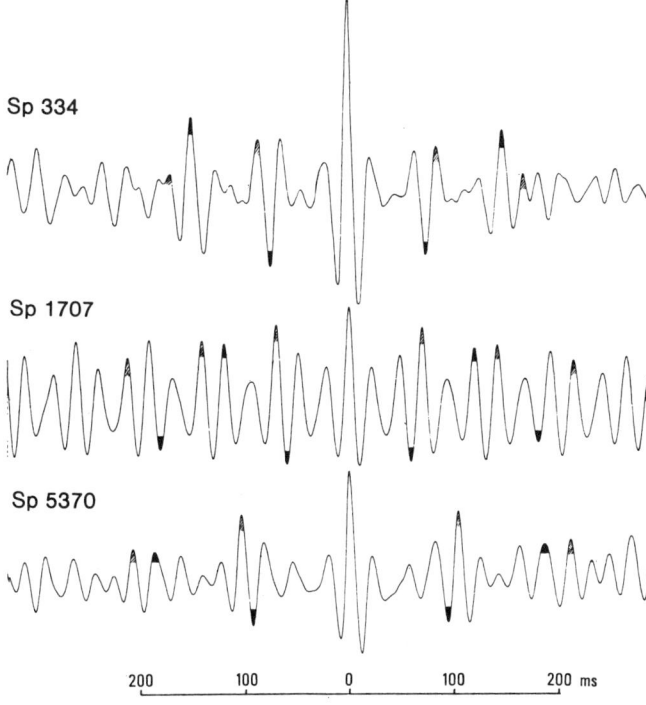

Figure 71.   Study of reverberation signs.

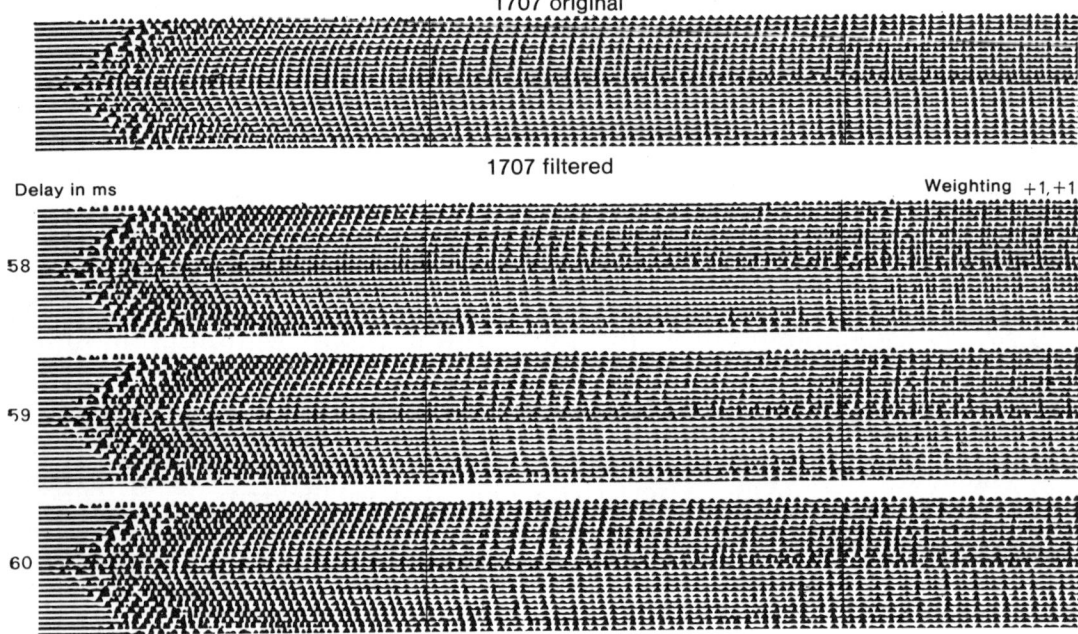

Figure 72.   Study of delays on shot 1707, reverberations of alternating signs.
Recording bandpass 18-75 hz.

This filter has been applied to SP 5370. Results are shown in Figures 74a and 74b. The two sign changes have been marked on the autocorrelation diagram. The filter designed to exclude reverberations of inverse sign, is seen to accept those of constant sign, although attenuated, and conversely. Finally, the lowest auto-correlation (Figure 74b) corresponding to the doublt filter no longer displays the trace of either category of reverberations.

Influence of weighting

The different sections that have been filtered were constructed by assuming an apparent reflection coefficient from 0.9 to 1; that is, in the two sample operator employed, the delayed trace had an amplitude equal to 90 or 100 percent that of the original trace.

This apparent reflection coefficient of 1 arises largely from the AGC action which reduces the natural attenuation of the multiples and therefore tends to maintain the periodic phenomenon.

5370 original

5370 filtered

Delay in ms

Weighting +1,−1

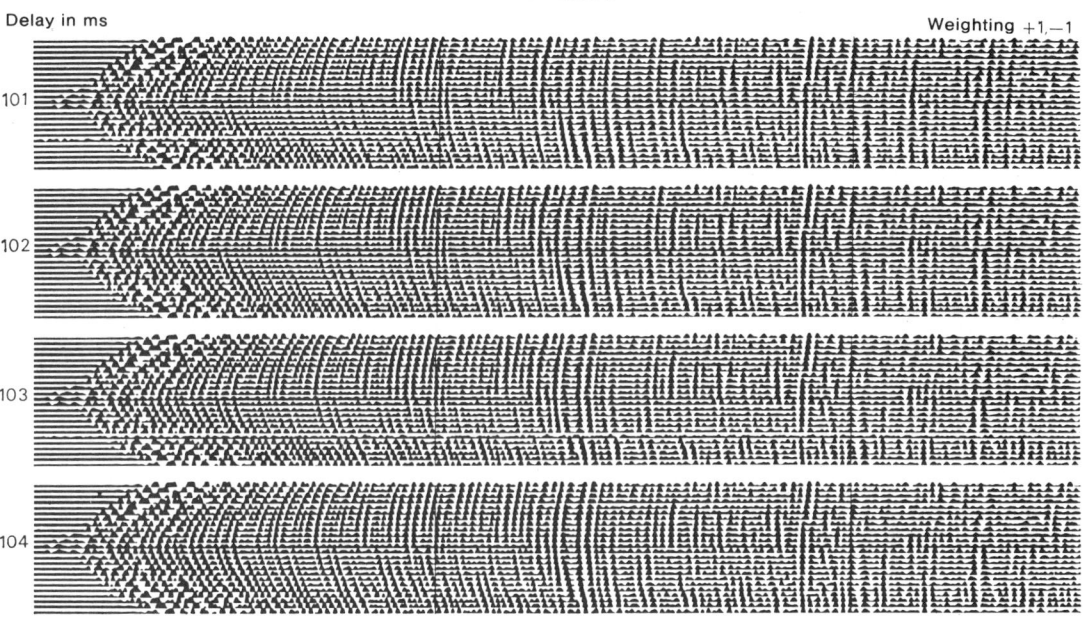

101

102

103

104

Figure 73.   Study of delays on shot 5370, reverberations with the same sign.
Recording bandpass 18-92 hz.

Tests conducted with several weightings on two shots, differing in location and quality, show the development of filtering as a function of the weighting (Figure 75a and 76a). The corresponding autocorrelations (Figure 75b and 76b) further emphasize this development in showing the progressive attenuation of multiples as the weights tend to become equal in absolute value.

If this development is interpreted in the form of attenuation is a function of frequency, it is seen that the frequencies are attenuated more and more in order to attain 45 db with equal weighting.

This means that of the two parameters which enter into the operator, namely the weighting value and the period in milliseconds, we can practically eliminate the first. This allows us to define the second parameter with all the desired accuracy. In a more general manner, we have seen that a very small error in the period is capable of rendering the filter ineffective, however elaborate it may be

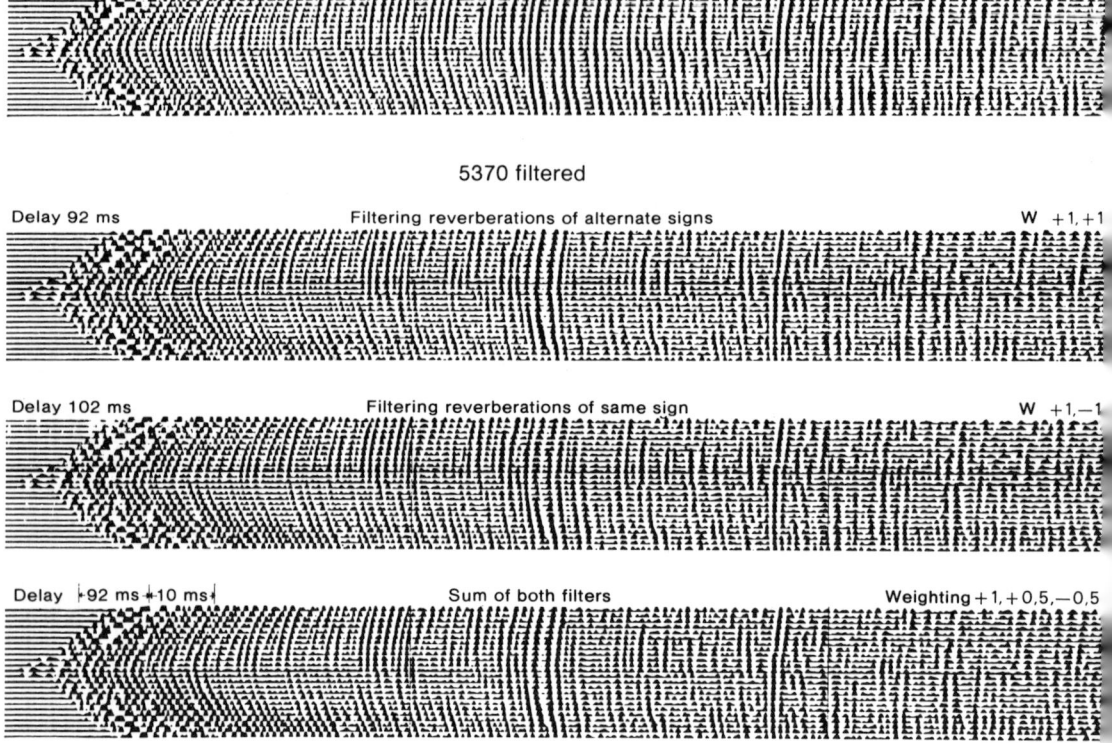

Figure 74a.   Effect of various filters on shot 5370. Recording filter 18-92 hz.

It is certain that in having neglected this accuracy, numerous attempts at filtering in isolated cases or in entire studies have been doomed to failure or have given only irregular results.

Study of first breaks. Value of reflection coefficient (r) at sea bottom

The study of first breaks should enable us to find a value of the water bottom reflection coefficient undistorted by the AGC action. Indeed, by considering that the first arrivals are produced by refraction at the sea bottom, the velocity can be found and by estimating the density, a sufficiently close value of the true r can be obtained. In very favorable cases, the calculation of the intercept should likewise be able to give us an idea of the refractor depth.

The study undertaken, on approximately 300 shots shows that these principles are confirmed wherever the traveltime curves can be plotted. Very often the distinction is not straightforward; nevertheless by considering an alignment on sev-

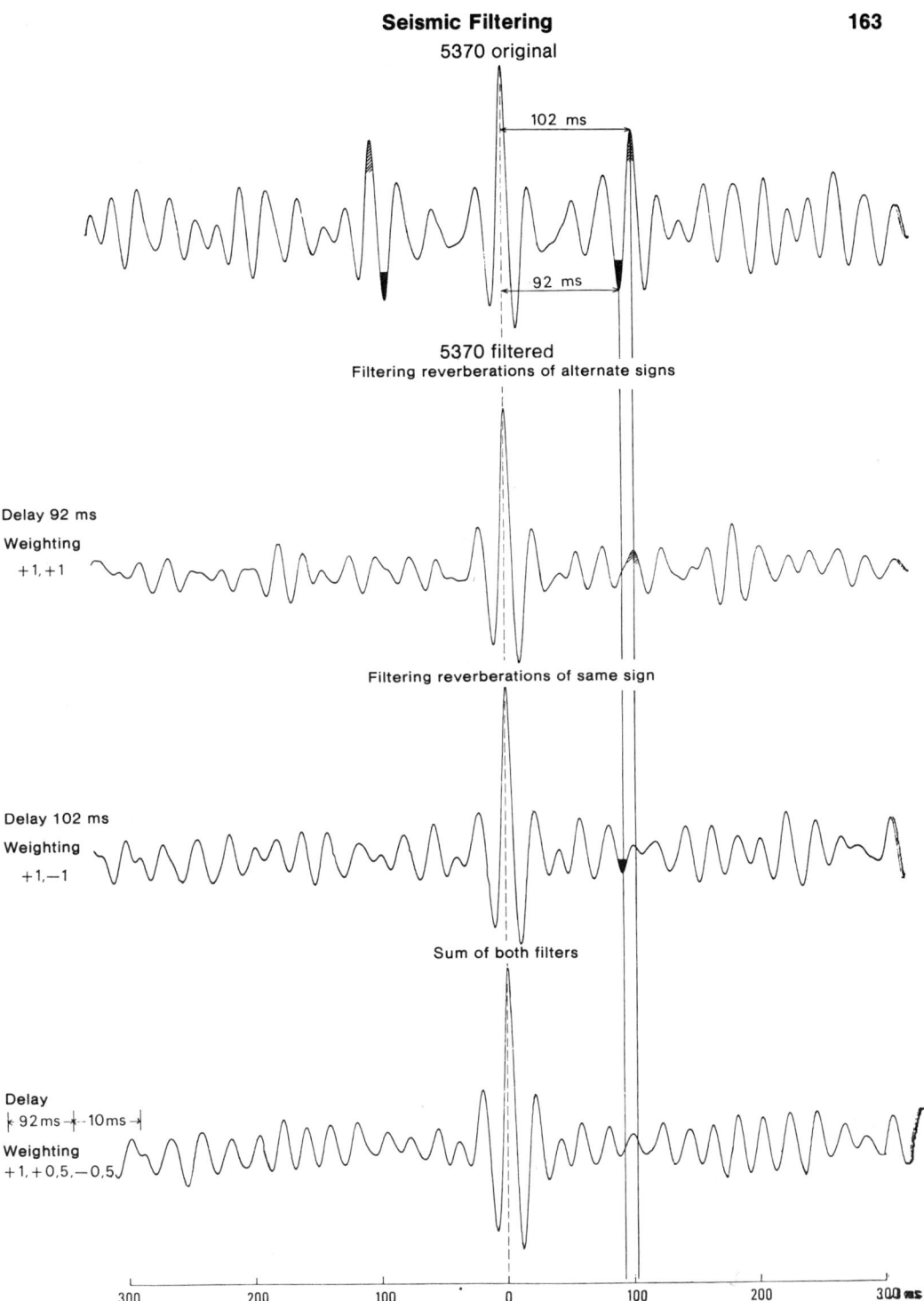

5370 original

102 ms

92 ms

5370 filtered
Filtering reverberations of alternate signs

Delay 92 ms

Weighting

+1,+1

Filtering reverberations of same sign

Delay 102 ms

Weighting

+1,−1

Sum of both filters

Delay

92 ms — 10 ms

Weighting
+1,+0,5,−0,5

300   200   100   0   100   200   300 ms

Figure 74b.   Autocorrelation of filtered traces from Figure 69a.

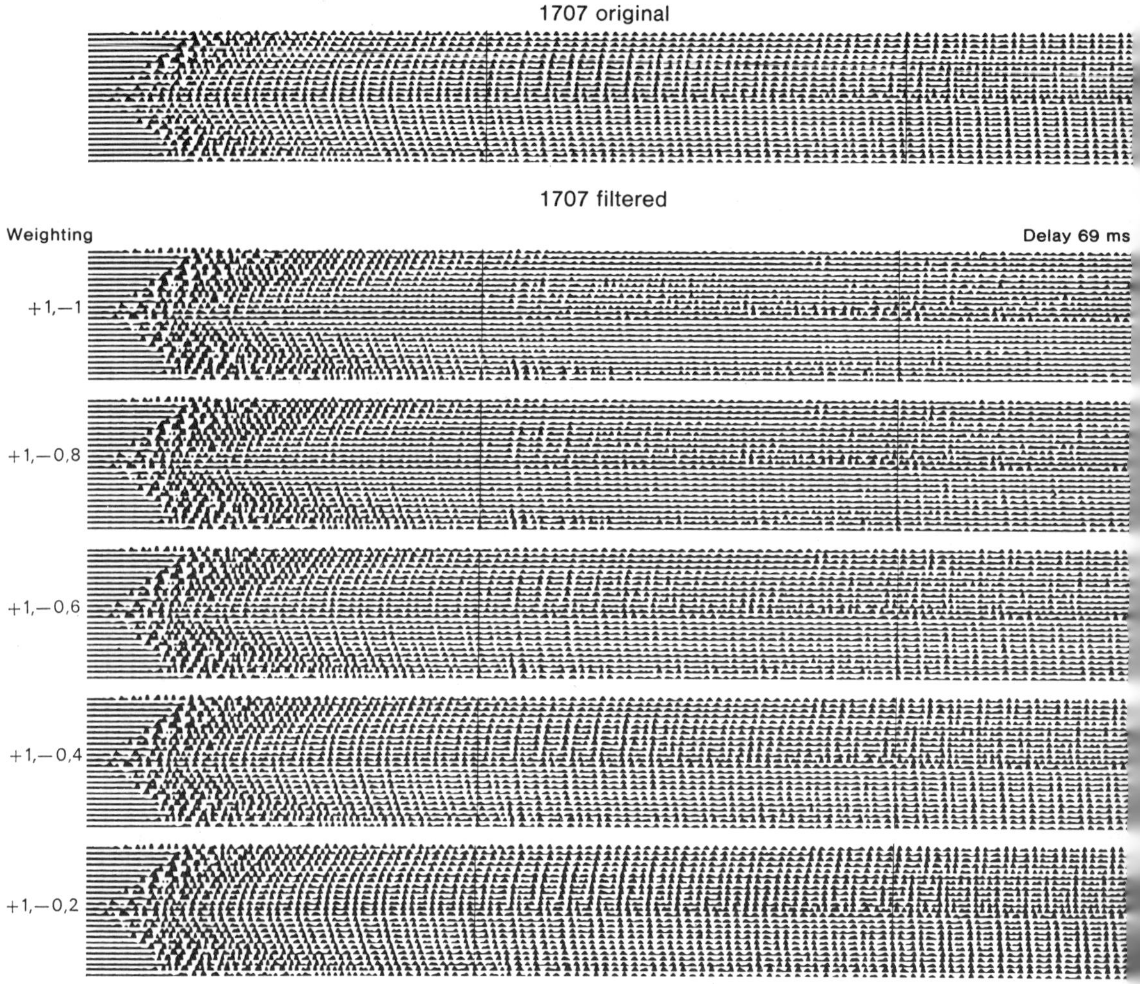

1707 original

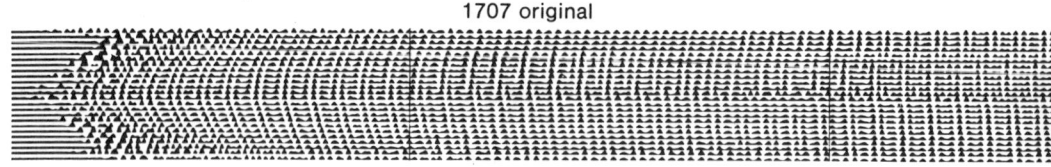

1707 filtered

Weighting   Delay 69 ms

+1,—1

+1,—0,8

+1,—0,6

+1,—0,4

+1,—0,2

Figure 75a.   Effect of weighting on record from shot 1707, reverberations
having the same sign. Recording filter 18-75 hz.

Autocorrelation of original record from Sp. 1707

Delay 69 ms

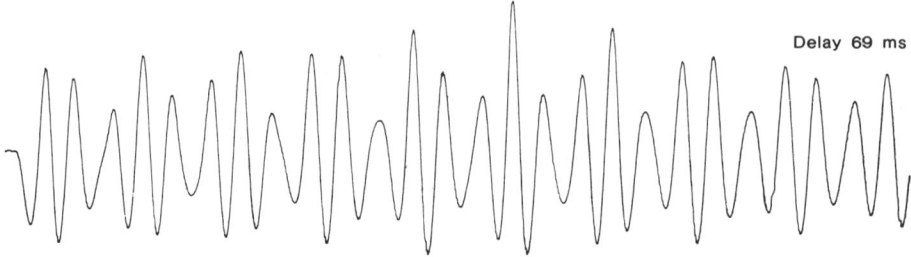

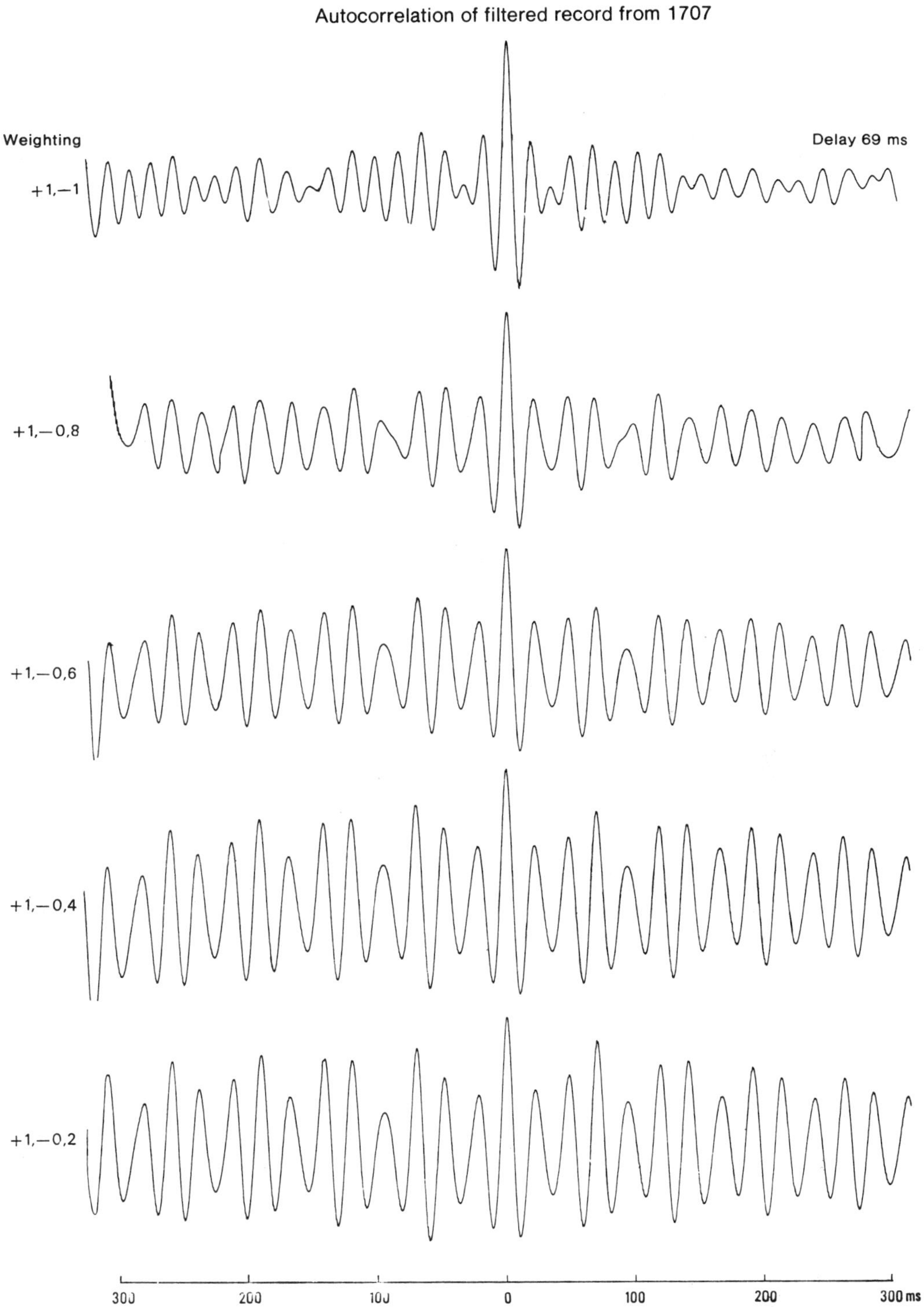

Autocorrelation of filtered record from 1707

Weighting

+1,−1

+1,−0,8

+1,−0,6

+1,−0,4

+1,−0,2

Delay 69 ms

300   200   100   0   100   200   300 ms

Figure 75b.   Autocorrelations of filtered traces from Figure 70a.

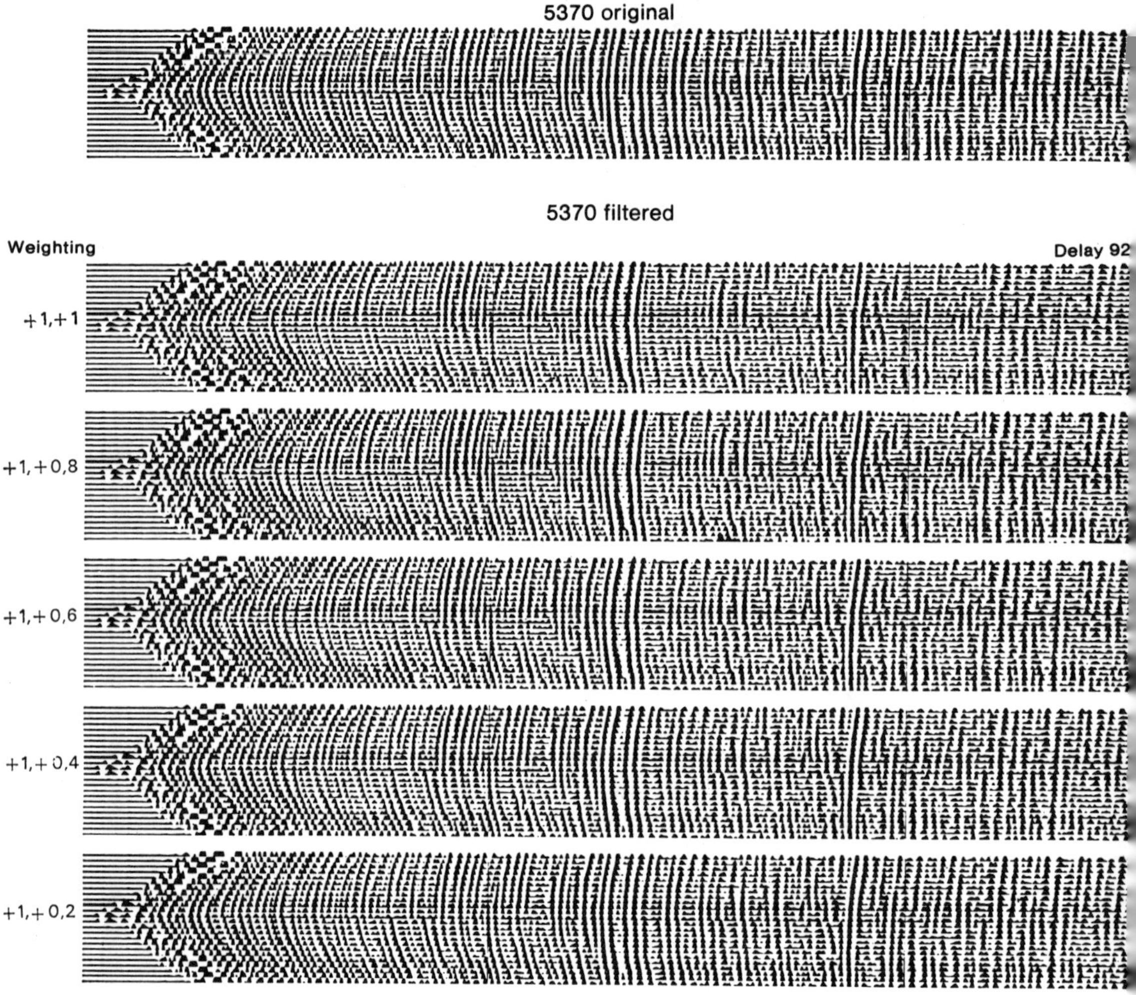

5370 original

Figure 76a. Effect of weighting on record from shot 5370, reverberations having alternating signs. Recording filter 18-75 hz.

Autocorrelation of original record from 5370

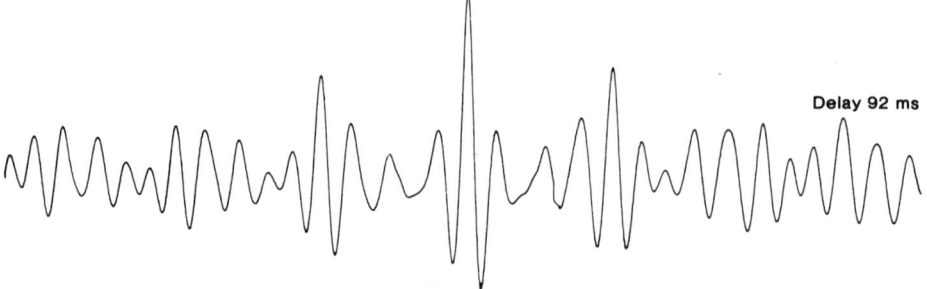

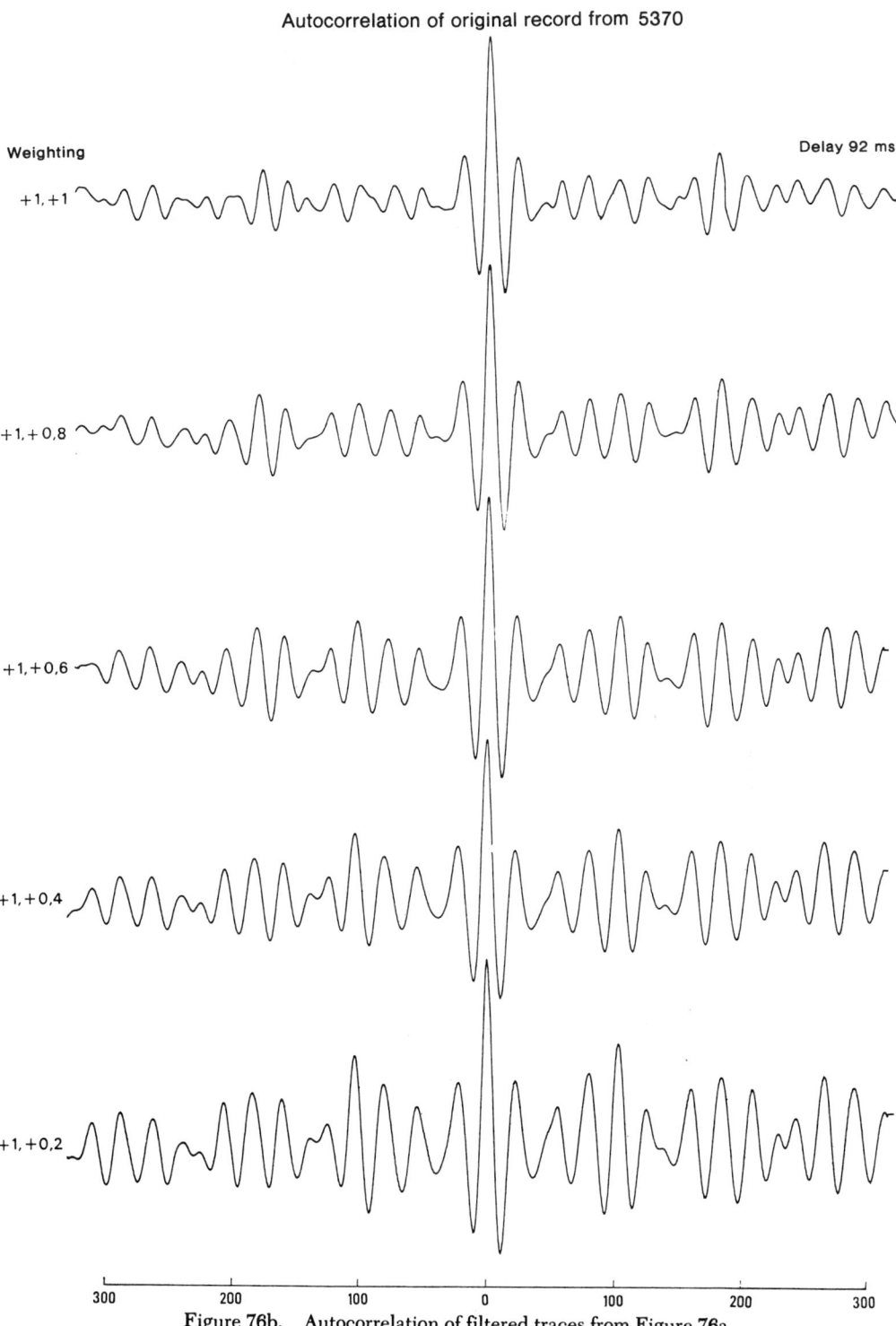

Figure 76b.   Autocorrelation of filtered traces from Figure 76a.

eral traces rather than an exact break on each trace, we can estimate whether the first arrivals have a velocity approaching that of water or not. Moreover, the accuracy with which the shot to detector distances are known is purely theoretical. The study pertains to shooting in the North Sea and off the coast of Gabon.

Gabon

In the example of singing (Figure 61a), where the water depth is less than 10m, the velocity found is 1450 m/sec.

On records 5370 and 5375-6, where the water depth is 65m, the velocities found are respectively 1610, 1550, and 1525; i.e., a velocity comparable to that of sound in water.

For 10 shots of another profile in the same area where the depth is about 10 m, we find:

1660	1550	1600	1550	1630
1700	1600	1620	1650	1710

i.e., an average V of 1627 m/sec.

North Sea

For the 19 shots of Figure 60a (reverberations), where the water depth is around 45 m, we find:

1750	1740	1700	1700	1715
1700	1740	1715	1700	1750
1730	1700	1700	1700	1760
1760	1700	1740	1700	

i.e., an average V of 1715 m/sec.

Shotpoint 334 (Figure 71), depth 54m.

The traveltime curve gives two velocities here: 1360 then 3190, which corresponds to a rather rare case. The first velocity obtained on traces closest to the shot is obviously spurious, since it pertains to the arrival of the direct wave in water with $V = 1500$, the error arises from the entirely theoretical shot-detector, and trace-interval distances taken. The second velocity is confirmed on eight traces.

These figures tend to show that in general the velocity of the sea bottom is of the same order of magnitude as that of the water. Confirmation of this is furnished by Suttonetal, 1957. The authors conducted very precise measurements on cores taken at the water bottom. These measurements pertain to the velocity of sound, bulk density, grain density, porosity, grain size, carbonate and salt content, etc.

The velocities recorded (231 measurements on 26 cores of basically different lithologies) are:

clays:	mean V 1550 (1430 to 1700)
clay silts, sands:	mean V 1650 (1600 to 1700)
calcarenites:	mean V 1700 (1650 to 2770)

Given these values, the actual coefficient at the sea bottom is therefore to a large extent due to the density contrast. In the article cited above, the density measurements ranged from 1.46 to 1.91. Densities and velocities vary broadly in the same sense with the following correspondence:

$$1500 \text{ m/sec}: d = 1.6$$
$$1600 \text{ m/sec}: d = 1.75$$
$$1700 \text{ m/sec}: d = 1.9$$

If r is calculated by means of the formula:

$$r = \frac{V_2 d_2 - V_1 d_1}{V_2 d_2 + V_1 d_1},$$

with

$$V_1 = V_2 = 1500 \text{ m/sec}$$
$$d_1 = 1$$
$$d_2 = 1.6,$$

we obtain: $r = 0.235$, the sea bottom reflection coefficient value which can be considered as minimal.

This is the case for shotpoints 5370-5-6 in Gabon as well as the shotpoints in the singing section of Figure 61a.

For the shots in the profile where $V_{2m} = 1627$ if $d_2 = 1.75$, $r = 0.31$.

The first example in the North Sea (Figure 60a) gives $V_{2m} = 1715$; i.e., with $d_2 = 1.9$, $r = 0.37$.

In the case of SP 334, where two velocities are found, it seems that the sea bottom constitutes a fast bed. Calculation of the water layer thickness by the intercept gives 60 m instead of the actual 54 m. This error is probable, owing to the small accuracy of the record picking and shot-detector distances, which leads us to obtain $r = 0.62$ if we take $d_2 = 2$.

This study of traveltime curves, as inaccurate as it may be, indicates two things:

1) The actual minimum reflection coefficient at the water bottom is of the order of 0.25.

2) Some high values, such as 0.8 mentioned in the Russian literature (5) or 0.6 found here, are not exceptional.

These actual values from 0.25 to 0.8 are less than those which appear on the records and which must be considered when filtering. The difference is due to the controlling action of the AGC which is interpreted by an apparent increase in the reflection coefficient whose new value can reach 0.9 or 1.

Comparisons of shots filtered by the two-sample operator and by the Backus filter. AGC action on (r).

Let us first specify the notations:

r = actual reflection coefficient at sea bottom

r_A = apparent reflection coefficient or weighting coefficient applied in the two-sample filter

r_B = reflection coefficient employed in the Backus formula

The Backus method has been applied at the two shotpoints 1707 and 5370, where we have previously studied the effect of weighting (Figures 75a to 76b).

The operator is written with consideration for the reverberation of alternating sign (case 5370):

$$F(t) = g(t) + 2\, r_B\, g(t-T) + r_B^2\, g(t-2\,T),$$

with $T =$ period reverberation.

If the reverberations are of the same sign (case of shotpoint 1707), it is necessary to change $2r_B$ to $-2r_B$ and replace T by T'. The parameter period being precisely known, we have made r_B vary from 0.1 to 1 and have displayed the records obtained, in Figures 77a and 78a and the corresponding autocorrelations in Figures 77b and 78b.

> SP 1707. Figure 77a displays records shot without changing the gain control.
> From $r_B = 0.1$ to $r_B = 0.5$, the reverberations are progressively attenuated;
> $r_B = 0.6$, beyond one second, only the reflected arrivals remain;
> from $r_B = 0.7$ to $r_B = 1$, the reflections are attenuated in turn.

Figure 77b represents the corresponding autocorrelations recorded with a gain such that the central peaks are essentially of the same amplitude, from $r_B = 0.1$ to 0.5 the reverberations attenuate placing the peak into progressively greater prominence; when $r_B = 0.6$, the reverberation closest to the peak is reduced to the main level of the trace, the others are suppressed; from $r_B = 0.7$ to 1.0, a new periodic phenomenon of different period appears and and is amplified in approaching $r_B = 1$.

From examination of these two figures, it follows that only $r_B = 0.6$ is suitable for filtering shot 1707 by the Backus method. Indeed, with $r_B = 0.5$ the reverberations still exist on the record, and commencing with $r_B = 0.7$ new reverberations appear clearly visible on the autocorrelations.

This result of filtering is comparable to that obtained for the same record with the two-sample operator having equal coefficients (Figure 75a and 75b). The records are identical except for the gain levels. The autocorrelations display the same frequency, and if the variations of character do not behave in the same sense, they occur at the same time. Therefore, r_A may be taken to equal 1.

In other respects, we note that there is a similarity between the autocorrelations:
$r_B = 0.4$ and $r_A = 0.8$;
$r_B = 0.3$ and $r_A = 0.6$.

SP 5370. Figure 77a displays the records shot without changing the gain control.

From $r_B = 0.1$ to 0.3, the reverberations progressively attenuate; as this record is less disturbed than SP 1707, let us consider the first reverberation from the 1800-msec horizon, occurring on the original a little before the 2-sec mark;
$r_B = 0.4$ and $r_B = 0.5$, the reverberations have disappeared;
from $r_B = 0.6$ to 1, new reverberations appear, notably the one before the 2-sec mark.

Figure 78b represents the corresponding autocorrelations recorded with a gain such that all the central peaks are perceptibly the same amplitude.

From $r_B = 0.1$ to 0.3, the autocorrelations represent in progressively attenuated form, the same character as the original autocorrelation;

from $r_B = 0.4$ to 0.6, the reverberation closest to the peak has disappeared, only the second reverberation remains in attenuated form;

from $r_B = 0.7$ to 1, there is reappearance of the first reverberation, as was evident on the records.

Definition of the better filtered record is more subtle here since the reverberation phenomenon is less marked than at SP 1707.

Among the r_B values of 0.4, 0.5, 0.6, the best autocorrelation seems to be $r_B = 0.5$, for it displays the smallest amplitude ratio between the second repetition and the autocorrelation peak.

From the records, $r_B = 0.4$ permits prediction of the reverberation located a little before 2 sec, whereas it has totally disappeared on $r_B = 0.5$. On the other hand, with $r_B = 0.5$, the filtering seems to be less effective beyond 2 sec. The best r_B to employ therefore seems to be between 0.4 and 0.5, i.e., $r_B = 0.45$.

Compared with the result obtained on this shot by the two-sample operator (Figure 76a and 76b), the records are noted to be identical, pair for pair: $r_B = 0.5$ and $r_A = 1$; $r_B = 0.4$ and $r_A = 0.8$; $r_B = 0.3$ and $r_A = 0.6$; $r_B = 0.2$ and $r_A = 0.4$.

Given this identity, the same conclusion holds for the subject of the best record and the r_A value ought to be between 0.8 and 1, i.e., 0.9.

As to the autocorrelations, although those concerning the weighting study are more enlarged, it is seen that they are completely identical pair for pair, according to the tabulation above.

The conclusion that emerges from this comparison of two methods of filtering, is as follows. With exact knowledge of the period, the Backus filter gives a very good result, if we are able to know r_B to within ± 0.05. The two-sample filter gives an almost identical result, or at least entirely valid, by employing $r_A = 1$. This value can drop to 0.9 in the case (type 5370) where the disturbing phenomenon does not totally mask the seismic arrivals.

In other respects, we observe:

1) that the apparent reflection coefficient r_A, employed for the two-sample filter, derived from the actual coefficient r after modification by the AGC, has a value rapidly tending towards 1;

2) that the reflection coefficient r_B employed by Backus lies between r and r_A.

Utilizing the values of the actual reflection coefficient r found during the study of traveltime curves, we attempt to establish a relation between r, r_A, and r_B.

Let us recall that:

for 1707 : $r = 0.37$

for 5370 : $r = 0.23$.

The different values of r found for records 1707 and 5370 respectively are: $r = 0.37$ and 0.23, $r_B = 0.6$ and 0.45, $r_A = 1$ and 0.90.

1707 original

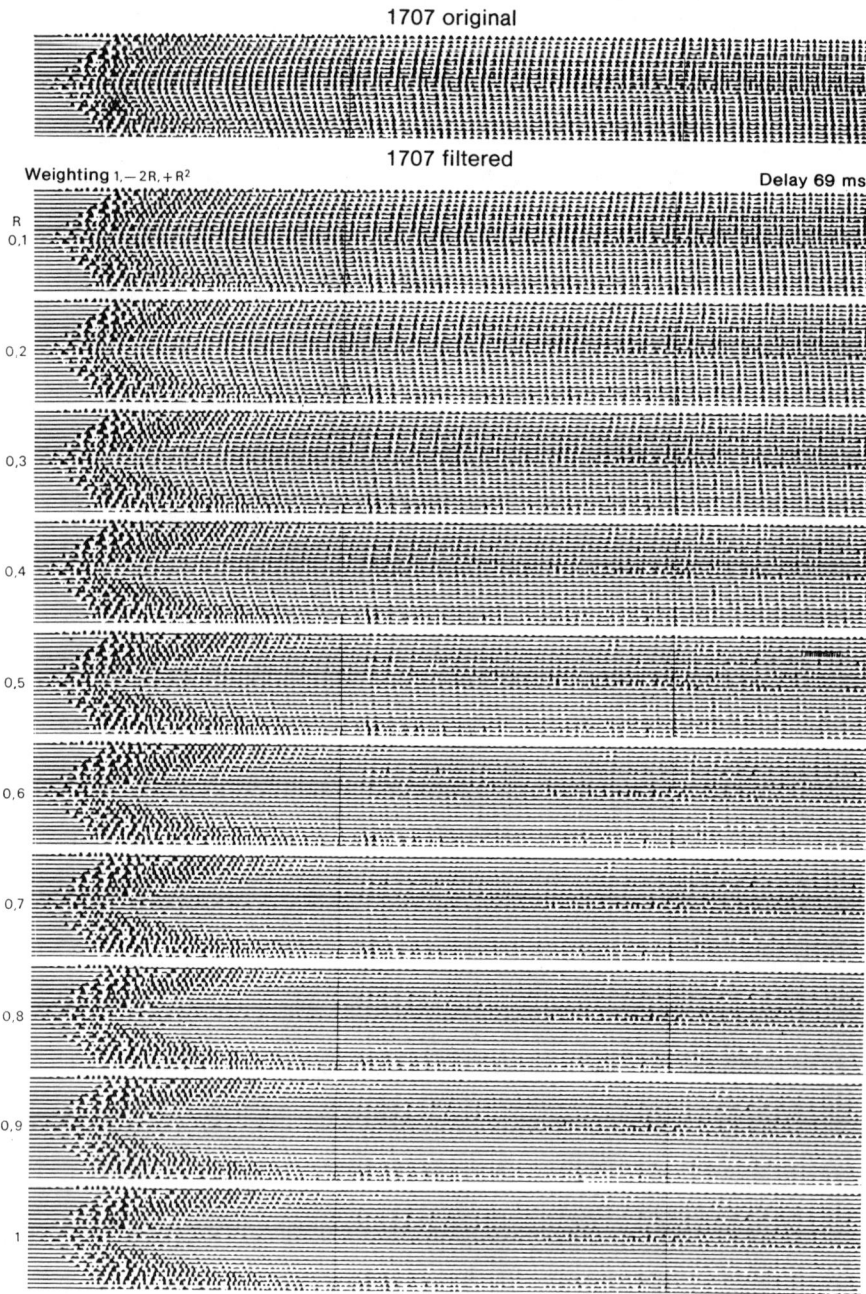

1707 filtered

Weighting $1, -2R, +R^2$ Delay 69 ms

R
0,1

0,2

0,3

0,4

0,5

0,6

0,7

0,8

0,9

1

Figure 77a. Filtered records from shot 1707, using the Backus formula, reverberations having the same sign. Recording filter 18-92 hz.

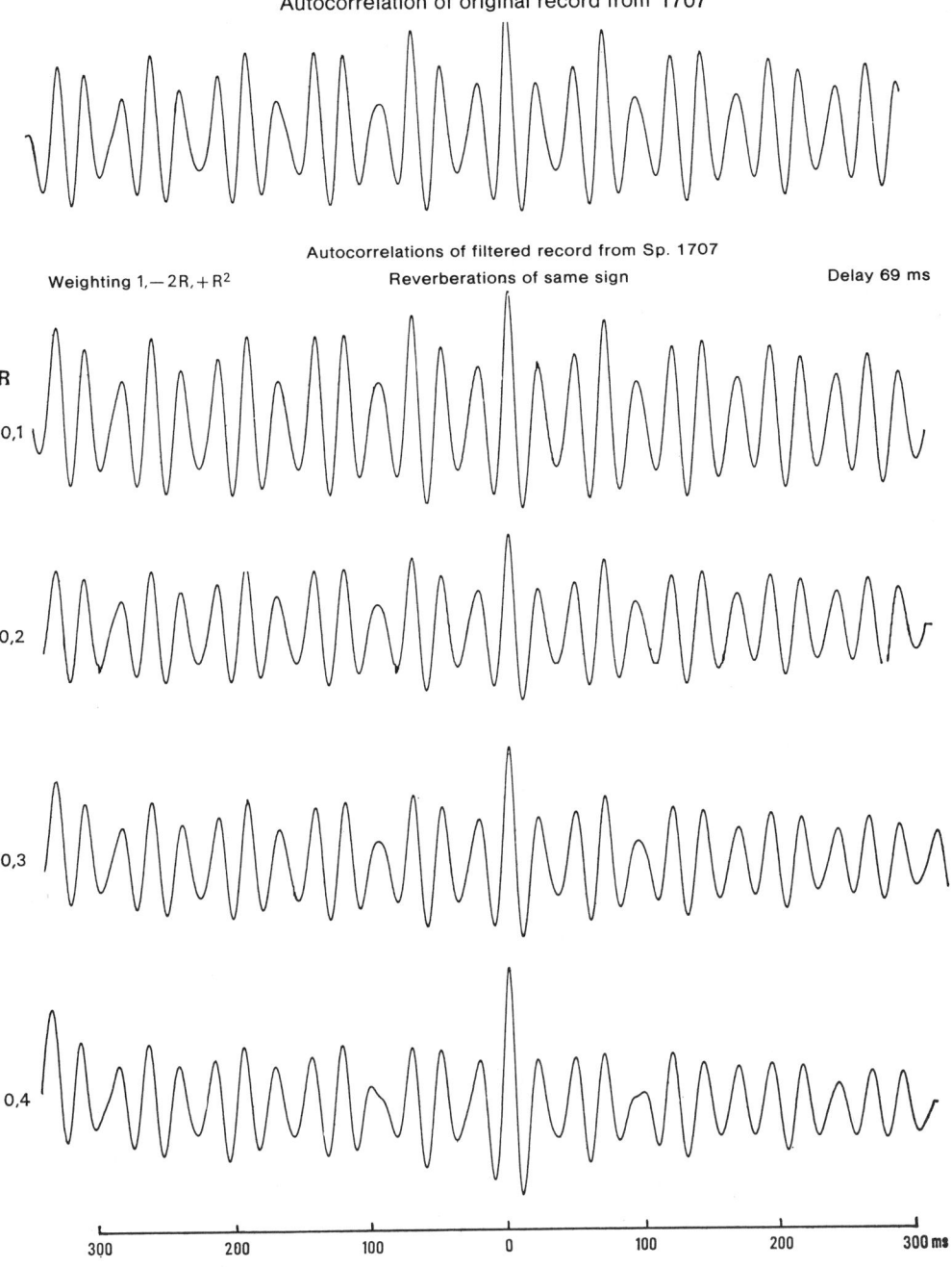

Autocorrelation of original record from 1707

Autocorrelations of filtered record from Sp. 1707

Weighting $1, -2R, +R^2$ Reverberations of same sign   Delay 69 ms

R

0,1

0,2

0,3

0,4

300 200 100 0 100 200 300 ms

Figure 77b.   Autocorrelations of filtered traces from Figure 77a.

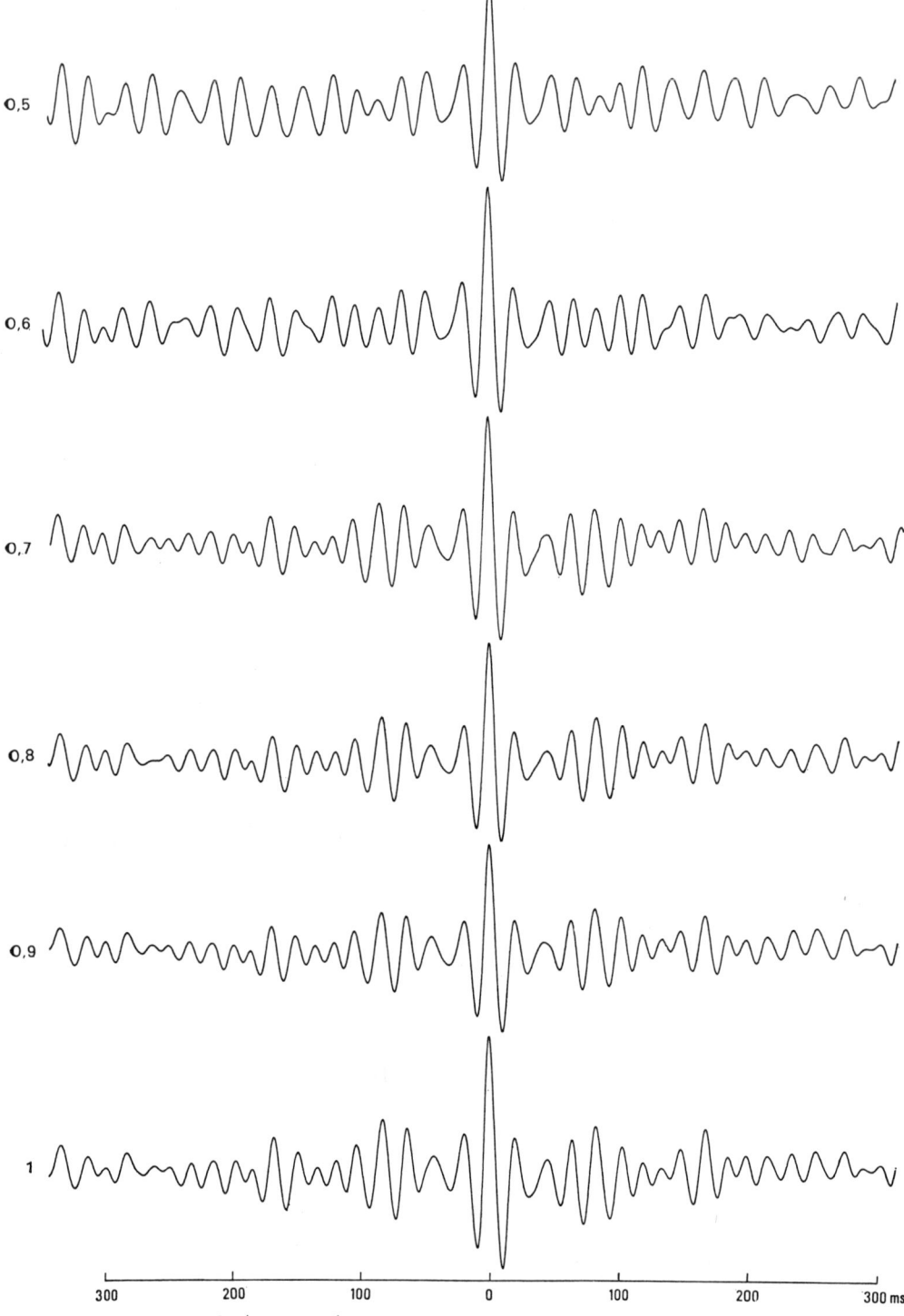

0,5

0,6

0,7

0,8

0,9

1

| 300 | 200 | 100 | 0 | 100 | 200 | 300 ms |

Figure 77b. (continued) Autocorrelations of filtered traces from Figure 77a.

If we consider the pair by pair relations of these coefficients we obtain:

$$1707: \frac{r_B}{r} = \frac{0.6}{0.37} \neq 1.6 \ . \ \frac{r_A}{r_B} = \frac{1}{0.6} \neq 1.6$$

$$5370: \frac{r_B}{r} = \frac{0.45}{0.23} \neq 2 \ . \ \frac{r_A}{r_B} = \frac{0.9}{0.45} \neq 2.$$

hence, for a given record:

$$\boxed{\frac{r_B}{r} = \frac{r_A}{r_B} = C}$$

This coefficient C is due to the AGC employed (In this case, amplifiers HTL 7000, AGC medium).

For use of the Backus filter:

For use of the two-sample filter:

$$\boxed{\begin{array}{l} r_B = Cr \\ r_A = Cr_B = C^2 r \end{array}}$$

and:

$$\boxed{C^2 = \frac{r_A}{r}}$$

This coefficient C is, on the other hand, a function of r. In fact, starting with a train of rapidly decaying arrivals (as is the case for a seismic arrival at the geophones preceding its train of reverberations) the AGC working in expansion, gives the same train of arrivals, but with a slower decay. Inspection of an AGC test suffices to observe its action. A sinusoidal signal of constant frequency is introduced into the AGC loops. The signal is caused to diminish in amplitude by increments of 10 db until 60 db. After a time of expansion depending on the instrumentation, the 60 db of input signal finds itself reduced to 6 db on output.

Adjustment of this expansion time can likewise play a part. Thus for the medium AGC employed in the 7000 B instrument, the expansion time is 60 msec; this time being less than the period of the reverberations. With a similar adjustment, it seems that the AGC would finish accommodating one arrival when the following one occurs. On the other hand, if this expansion time were longer (the wideband adjustment has $t_e = 230$ msec.), it seems that after a strong arrival, the amplification being weak, the first reverberation would be weakly amplified, and would permit enhancement of the useful arrival.

We have seen that the actual marine reflection coefficient was of the order of 0.3, that is, the reverberation of a seismic arrival is attenuated by 10 db with each reflection interference from the sea bottom. In other respects, the AGC reduces the variation of the input level in the ratio 10 to 1. The 10 db attenuation of the first reverberation with respect to the actual arrival is therefore going to be reduced to 1 db; i.e., an apparent r of 0.9. Moreover, if r could be equal to 1, r_A and r_B would likewise be equal to 1.

From these considerations, we can then construct the curve $r_A = f(r)$, $r_B = f(r)$, as well as the curve $C = f(r)$ (Figure 79).

Analog Filtering of Marine Seismic Records

5370 original

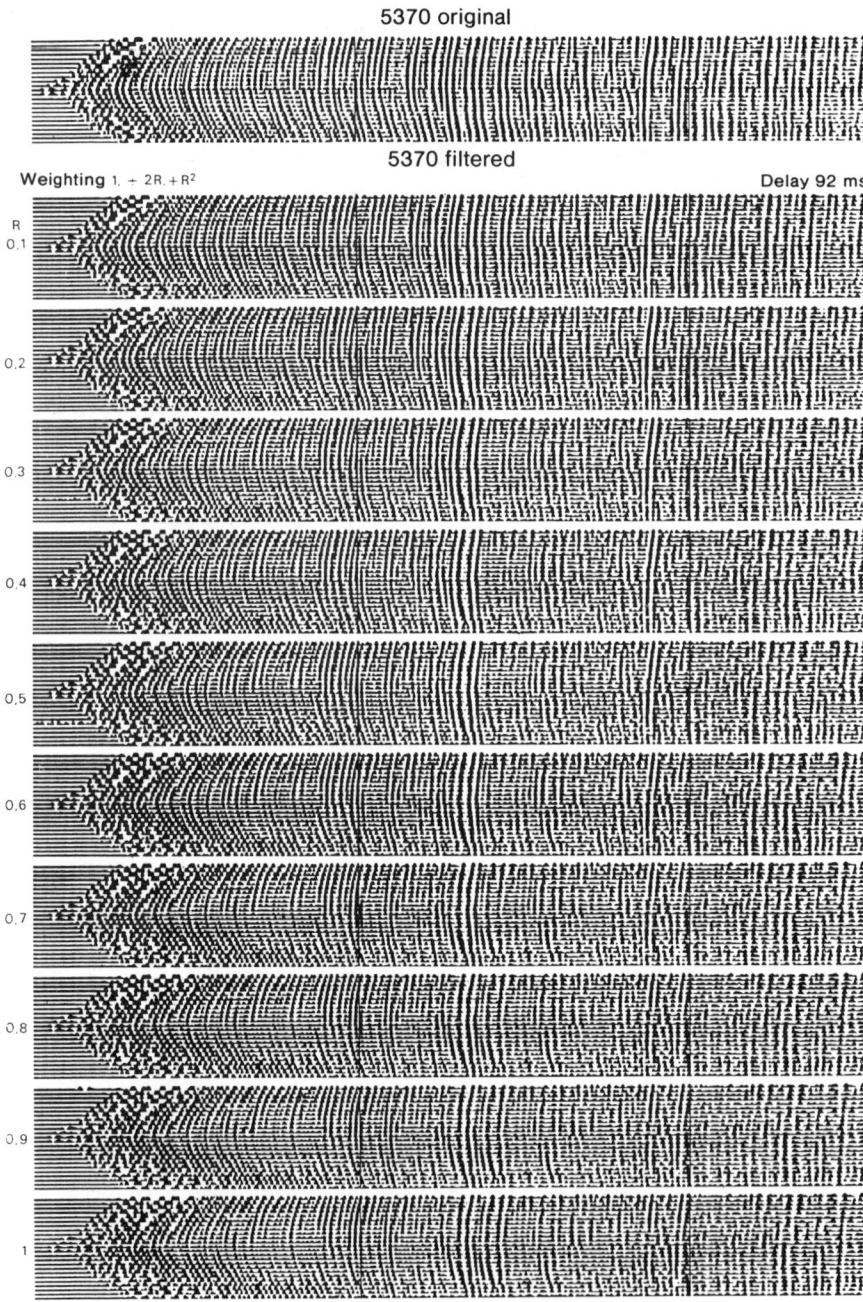

Figure 78a. Filtered records from shot 5370, using the Backus formula, reverberations having alternating signs. Recording filter 18-92 hz.

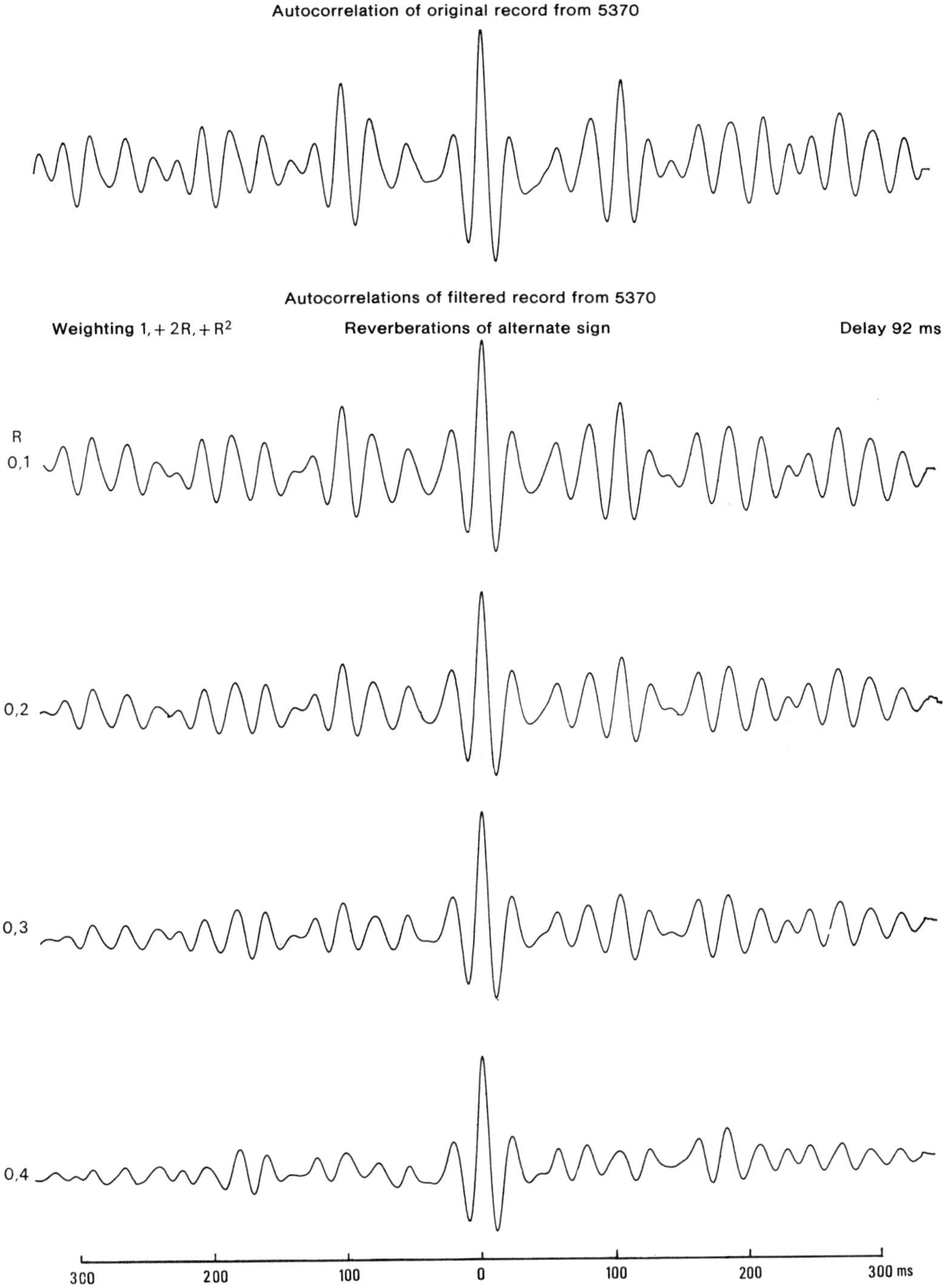

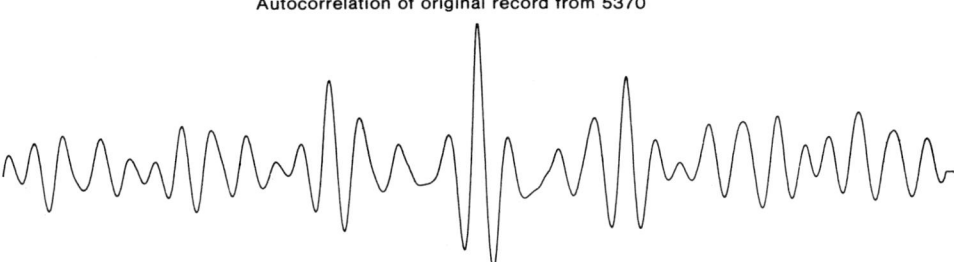

Autocorrelation of original record from 5370

Autocorrelations of filtered record from 5370

Weighting 1, + 2R, + R^2 Reverberations of alternate sign   Delay 92 ms

R
0,1

0,2

0,3

0,4

300   200   100   0   100   200   300 ms

Figure 78b.   Autocorrelations of filtered traces from Figure 78a.

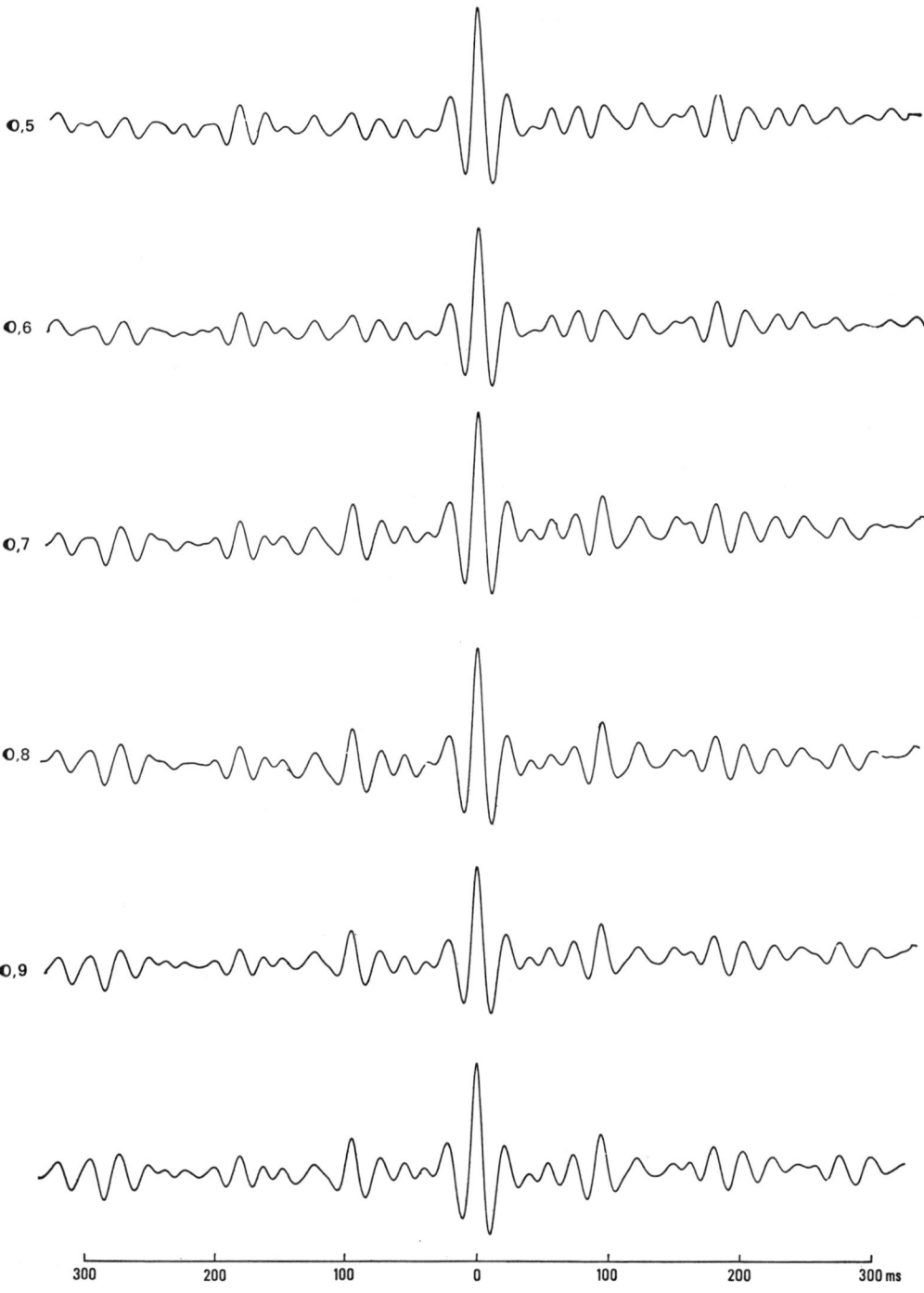

Figure 78b.   (continued).

Let us remember that these curves result from the use of a particular AGC (7000 B, Medium). Other types of AGC would give other curves.

We consider the actual values of r found for shotpoints 5370 and 1707 and, by way of confirmation, calculate from the curves of C, r_B, and r_A:

			ESTIMATED VALUES FROM RECORD FILMS AND AUTOCORRELATIONS
5 370. $r = 0.23$		$r_A = 0.86$	0.9
1 707. $r = 0.37$		$r_A = 0.91$	1
$C^2 = \dfrac{r_A}{r}$	5 370 : $C = \sqrt{\dfrac{0.86}{0.23}} = 1.935$		2
	1 707 : $C = \sqrt{\dfrac{0.91}{0.37}} = 1.57$		1.6
$r_B = Cr$	5 370 : $r_B = 0.23 \times 1.935 = 0.445$		0.45
	1 707 : $r_B = 0.37 \times 1.57 = 0.58$		0.6

We now see why the two-sample filter is rather insensitive to variation of the weighting around unity; in fact, the minimum value of r, 0.23, is expressed by the AGC in an r_A of 0.86, and its maximum value, 0.8, is expressed by an r_A of 0.98. The variations of r_A therefore range from 0.86 to 1.

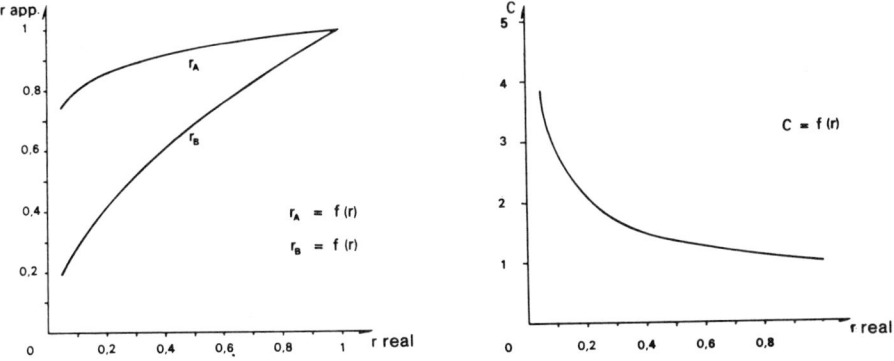

Figure 79.   Curves showing the action of AGC on the coefficient of reflection.

CONCLUSIONS

The conclusions which emerge from experiments conducted on records displaying reverberations solely, allow us to focus on a method or a course to follow in order to filter such records. This experimental method in marine seismic work can be applied, moreover, to land seismic surveys in order to eliminate multiples generated in the weathered zone or between beds.

We have seen first of all (Figure 62) the danger that attends the use of narrow electrical filters in favor of low frequencies (0-40, 15-28, 20-40, 30-50 hz) without even knowing if the sections studied deserve such treatment in the absence of being able to do better. The singing that appears on these sections is not always the work of nature.

For shots in water deeper than 30m, the procedure of considering twice the bottom sounder depth, converting the time with $V = 1500$ m/sec and taking reflection coefficients of different sign at the top and bottom of the water layer, with a view to applying a filter more elaborate than the narrow electrical one, does not always guarantee good results. Indeed, even if this elementary approach were generally true, we have seen that the accuracy of the sounder is inadequate compared to the $\pm$ 0.5 msec precision required to implement a filter based on the two-way traveltime in the water layer.

Besides, the reflection coefficient at sea bottom may be negative, giving rise to reverberations of constant sign (Figure 63, bottom shot) corresponding to the water depth. Furthermore, the time difference between two reverberations can not correspond to the water depth, but is greater, perhaps in the order of 10 msec because of the impulse shape, (SP 1707), or much more if a deeper horizon is involved (Figure 70). In these last two cases, the reverberations may be of constant or opposite sign, or sometimes present a hybrid phenomenon (5370).

We have later seen that for each filter which depends on the period of the reverberation, this period must be known with great accuracy, hence the need for frequent adjustment of the filter, on occasion even several times per shot. Constant results are thus obtained which cannot be attained with the one filter applied to an entire series of shots, then strongly modifying the filter to treat the subsequent series.

According to the type of filter employed, the reflection coefficient r to be considered is not the same. For the two-sample filter, it is the apparent r whose relation to the actual r is a function of the AGC employed. We should not hesitate to take this apparent r equal to 1, contrary to all physical probability, the AGC contributing to make the apparent r fairly constant. This is what facilitates filtering since only the period remains to be specified.

Assuming the same accuracy requirements for the period exists, the Backus filter is in other respects more sensitive to an error in the value of r which must be applied to it. In this regard it is more difficult to implement for results comparable to those obtained with the two-sample filter.

From these conclusions, a general method of filtering can be evolved which rests solely on the processing of the magnetic tape record.

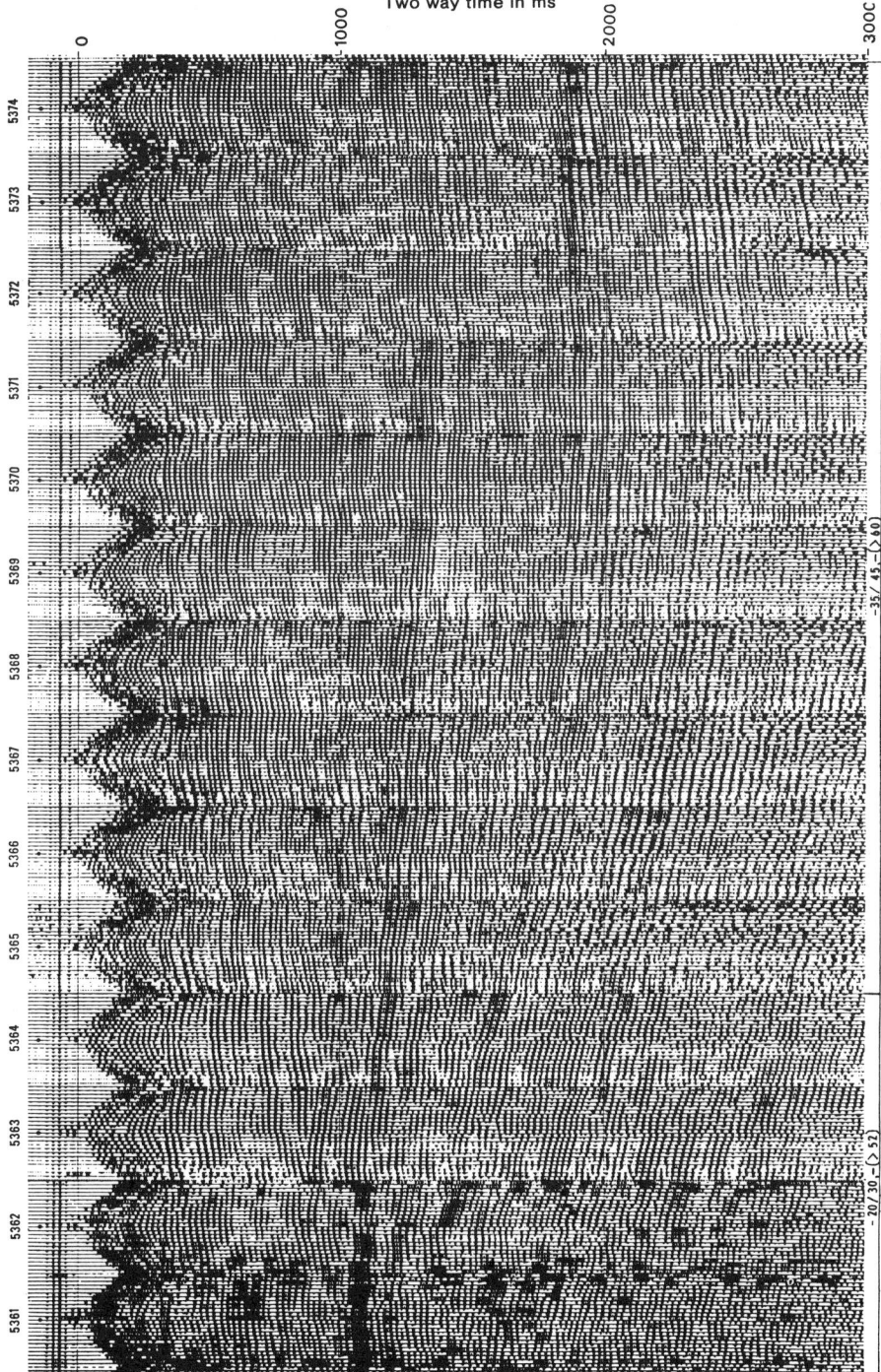

Figure 80.   Example of filtering done without examination of the original record.

Two way time in ms

Figure 81. Original records corresponding to Figure 80. Recording bandpass 18-92 hz.

Two way time in ms

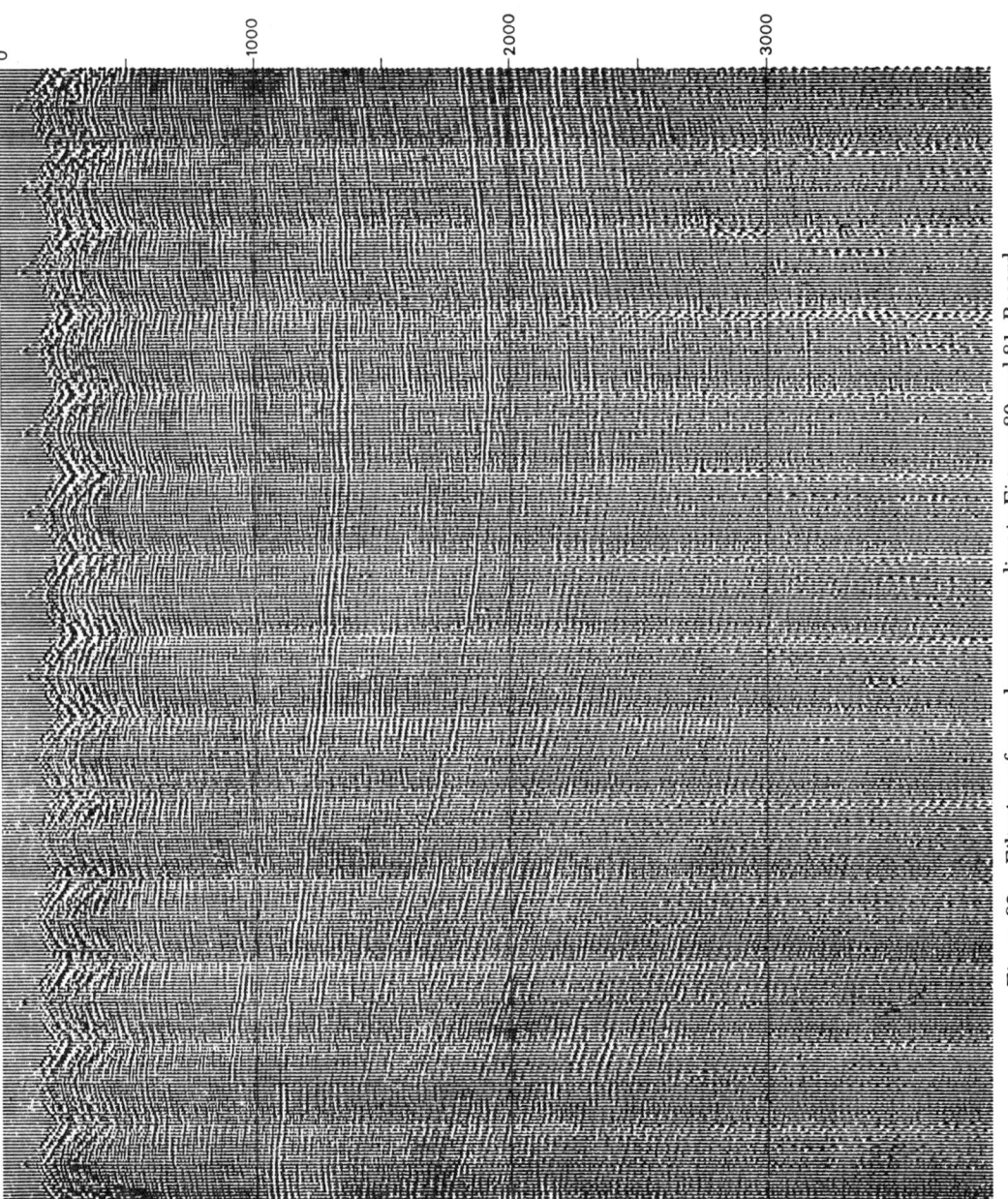

Figure 82. Filtering of records corresponding to Figures 80 and 81. Recording filter is 18 — 92 hz..

1. Study of the original sections with wide-band electrical filter. Selections of portions to be filtered since some zones can certainly be interpreted directly.

2. Application and study of autocorrelations for the selected portion. Analysis of interference and measurement of its parameters.

3. Application of an adequate filter, adjusted when necessary, on records uncorrected for normal moveout.

Refined sections will thus be obtained, on which it will be simple to establish velocity relations from actual arrivals.

As a final example, we consider Figures 80 and 82 which illustrate a procedure which is not recommended. Figure 80 represents a marine section where the water layer thickness increases from 60 to 70 m. It has been directly processed by the delay-line filter, without playback of the original records. At shotpoints 5361 to 5364, frequencies from 20 to 30 hz and above 52 hz have been attenuated; then from shotpoint 5365 to 5374, the attenuated frequencies are 35 to 45 hz and above 60 hz, the whole section being electrically filtered with 18-66 hz.

The quality of such processing cannot be judged for lack of the original data. The character change is clear between shotpoints 5364 and 5365, where the filter is modified.

A playback of the original recordings was made at a later date with an 18-92 hz electrical filter; it is displayed as Figure 81. It is seen to be interpretable, despite the reverberations.

Finally, the section was processed through a two- or four-sample filter, according to the shotpoint, and the result is displayed in Figure 82.

APPENDIX

CALCULATION OF Δt DIFFERENCE BETWEEN AN ACTUAL HORIZON AND ITS FIRST MULTIPLE

We consider a marker M located at a depth D (Figure 83) below a water layer of thickness Z, in which the velocity of sound is $v = 1500$ m/sec. The average velocity between the water bottom and M is V.

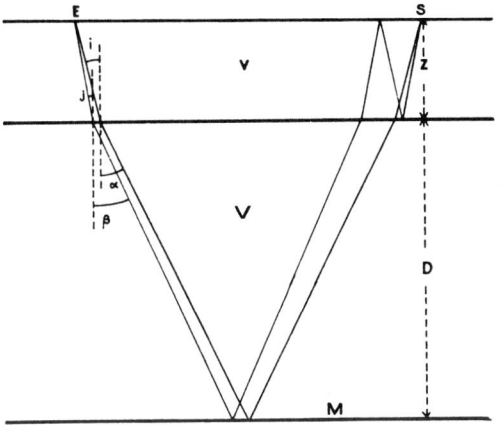

Figure 83.   Raypaths through an earth overlain by water.

An explosion at E at the time $t = 0$ generates a disturbance which propagates in the water layer, then in the substratum where after reflection from M, it returns to the water layer and again crosses it. It is then picked up at the time t_{1S} by the geophone S situated at the distance X from E (Figure 84).

After reflection at the air-water interface, the disturbance descends again to be reflected at the water bottom and returns to activate the geophone S at the time t_2 and so on until the energy is too weak to be seen.

We denote i and j, relating to t_1 and t_2, as the angles of incidence at the water-substratum interface, and α and β as the corresponding refraction angles in the substratum.

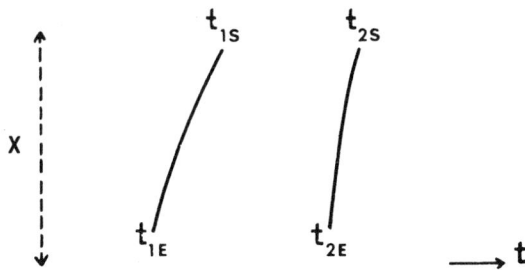

Figure 84.   Typical stepout curves.

We are going to express the Δt difference that exists between Δt_1, corresponding to the arrival of the M marker, and Δt_2 corresponding to the first multiple due to the water layer.

The arrival times at the geophone located close to the shot are t_{1E} and t_{2E}; and t_{1S} and t_{2S} are the arrival times at the geophone S located at the distance X from E,   So

$$\delta\Delta t = \Delta t_1 - \Delta t_2, \tag{1}$$
$$\Delta t_1 = t_{1S} - t_{1E},$$
$$\Delta t_2 = t_{2S} - t_{2E},$$

hence:

$$\delta\Delta t = t_{1S} - t_{2S} + t_{2E} - t_{1E}. \tag{2}$$

We have, as a function of the angles i, α and j, β

$$\left.\begin{array}{ll} t_{1S} = \dfrac{2Z}{v\cos i} + \dfrac{2D}{V\cos\alpha} & t_{1E} = \dfrac{2Z}{v} + \dfrac{2D}{V} \\[3mm] t_{2S} = \dfrac{4Z}{v\cos j} + \dfrac{2D}{V\cos\beta} & t_{2E} = \dfrac{4Z}{v} + \dfrac{2D}{V} \end{array}\right\} \tag{3}$$

Substituting in (2), we obtain:

$$\delta\Delta t = \frac{2Z}{v}\left(\frac{1}{\cos i} - \frac{2}{\cos j} + 1\right) + \frac{2D}{V}\left(\frac{1}{\cos\alpha} - \frac{1}{\cos\beta}\right). \tag{4}$$

Given that:

$$\frac{\sin i}{\sin\alpha} = \frac{v}{V}, \quad \sin\alpha = \frac{V}{v}\sin i$$

and

$$\cos\alpha = \sqrt{1 - \frac{V^2}{v^2}\sin^2 i} = \frac{\sqrt{v^2 - V^2\sin^2 i}}{v}$$

Likewise we have:

$$\cos\beta = \frac{\sqrt{v^2 - V^2\sin^2 j}}{v}$$

hence by substituting in (4)

$$\delta\Delta t = \frac{2Z}{v}\left(\frac{1}{\cos i} - \frac{2}{\cos j} + 1\right) + \frac{2D}{V}\left(\frac{v}{\sqrt{v^2 - V^2\sin^2 i}} - \frac{v}{\sqrt{v^2 - V^2\sin^2 j}}\right) \tag{5}$$

with

$$X = 2Z\tan i + \frac{2DV\sin i}{\sqrt{v^2 - V^2\sin^2 i}} = 4Z\tan j + \frac{2DV\sin j}{\sqrt{v^2 - V^2\sin^2 j}} \tag{6}$$

Equations (5) and (6) have been employed to calculate $\delta\Delta t$ as a function of X, V, t_1 and Z. The results for several sets of $\delta\Delta t$ values expressed in milliseconds, are recorded in the tabulations below.

TABULATION GIVING Δt DIFFERENCES BETWEEN ORIGINAL
HORIZON (Δt_1) AND FIRST MULTIPLE (Δt_2)

$$\delta \, \Delta t = \Delta t_1 - \Delta t_2$$

$$\boxed{X = 300 \text{ m}}$$

$V = 2\,000$ m/sec

Water Section

$Z =$	40 m	60 m	80 m	100 m	120 m	140 m	160 m	180 m	200 m
$t_1 = 500$ ms	1.17	1.69	2.17	2.62	3.04	3.41	3.75	4.04	4.28
750 ms	0.55	0.81	1.06	1.29	1.51	1.72	1.92	2.10	2.28
1 000 ms	0.32	0.47	0.61	0.75	0.89	1.02	1.14	1.26	1.38
1 250 ms	0.21	0.31	0.40	0.50	0.59	0.67	0.76	0.84	0.92
1 500 ms	0.15	0.21	0.28	0.35	0.41	0.48	0.54	0.60	0.66

$V = 2\,200$ m/sec

$Z =$	40 m	60 m	80 m	100 m	120 m	140 m	160 m	180 m	200 m
$t_1 = 500$ ms	0.81	1.17	1.50	1.81	2.08	2.33	2.53	2.70	2.84
750 ms	0.38	0.56	0.72	0.88	1.03	1.18	1.31	1.43	1.55
1 000 ms	0.22	0.32	0.42	0.52	0.61	0.70	0.78	0.86	0.94
1 250 ms	0.14	0.21	0.28	0.34	0.40	0.46	0.52	0.58	0.63
1 500 ms	0.10	0.15	0.19	0.24	0.28	0.33	0.37	0.41	0.45

$V = 2\,400$ m/sec

$Z =$	40 m	60 m	80 m	100 m	120 m	140 m	160 m	180 m	200 m
$t_1 = 500$ ms	0.57	0.83	1.06	1.27	1.46	1.62	1.76	1.86	1.93
750 ms	0.27	0.39	0.51	0.62	0.73	0.83	0.92	1.01	1.08
1 000 ms	0.15	0.23	0.30	0.37	0.43	0.49	0.55	0.61	0.66
1 250 ms	0.10	1.05	0.19	0.24	0.28	0.33	0.37	0.41	0.44
1 500 ms	0.07	0.10	0.14	0.17	0.20	0.23	0.26	0.29	0.32

$V = 2\ 600$ m/sec
Water Section

Z =	40 m	60 m	80 m	100 m	120 m	140 m	160 m	180 m	200 m
$t_1 = 500$ ms	0.42	0.60	0.77	0.92	1.05	1.16	1.25	1.31	1.35
750 ms	0.20	0.29	0.38	0.46	0.53	0.60	0.66	0.72	0.77
1 000 ms	0.11	0.17	0.22	0.26	0.31	0.36	0.40	0.44	0.47
1 250 ms	0.07	0.11	0.14	0.17	0.21	0.24	0.26	0.29	0.32
1 500 ms	0.05	0.08	0.10	0.12	0.15	0.17	0.19	0.21	0.23

$V = 2\ 800$ m/sec

Z =	40 m	60 m	80 m	100 m	120 m	140 m	160 m	180 m	200 m
$t_1 = 500$ ms	0.31	0.45	0.57	0.68	0.77	0.84	0.90	0.94	0.96
750 ms	0.14	0.21	0.27	0.33	0.39	0.44	0.49	0.53	0.56
1 000 ms	0.08	0.12	0.16	0.20	0.23	0.26	0.29	0.32	0.35
1 250 ms	0.05	0.08	0.10	0.13	0.15	0.17	0.20	0.22	0.24
1 500 ms	0.04	0.06	0.07	0.09	0.11	0.12	0.14	0.15	0.17

$V = 3\ 000$ m/sec

Z =	40 m	60 m	80 m	100 m	120 m	140 m	160 m	180 m	200 m
$t_1 = 500$ ms	0.24	0.34	0.43	0.51	0.58	0.63	0.67	0.69	0.70
750 ms	0.11	0.16	0.21	0.25	0.29	0.33	0.36	0.39	0.42
1 000 ms	0.06	0.09	0.12	0.15	0.17	0.20	0.22	0.24	0.26
1 250 ms	0.04	0.06	0.08	0.10	0.11	0.13	0.15	0.16	0.18
1 500 ms	0.03	0.04	0.06	0.07	0.08	0.09	0.11	0.12	0.13

$$\boxed{X = 400 \text{ m}}$$

$V = 2\,000$ m/sec

Water Section

$Z =$	40 m	60 m	80 m	100 m	120 m	140 m	160 m	180 m	200 m
$t_1 = 500$ ms	1.97	2.87	3.70	4.46	5.16	5.79	6.35	6.85	7.26
750 ms	0.96	1.41	1.84	2.24	2.63	2.99	3.34	3.66	3.97
1 000 ms	0.56	0.83	1.08	1.33	1.56	1.79	2.01	2.22	2.43
1 250 ms	0.37	0.54	0.71	0.87	1.04	1.20	1.35	1.49	1.63
1 500 ms	0.26	0.38	0.50	0.62	0.73	0.84	0.95	1.06	1.16

$V = 2\,200$ m/sec

$Z =$	40 m	60 m	80 m	100 m	120 m	140 m	160 m	180 m	200 m
$t_1 = 500$ ms	1.37	1.98	2.54	3.06	3.52	3.93	4.30	4.60	4.82
750 ms	0.66	0.97	1.26	1.54	1.80	2.05	2.28	2.50	2.70
1 000 ms	0.38	0.57	0.74	0.91	1.07	1.23	1.38	1.52	1.66
1 250 ms	0.26	0.38	0.49	0.60	0.71	0.82	0.92	1.02	1.12
1 500 ms	0.18	0.26	0.34	0.42	0.50	0.58	0.65	0.73	0.80

$V = 2\,400$ m/sec

$Z =$	40 m	60 m	80 m	100 m	120 m	140 m	160 m	180 m	200 m
$t_1 = 500$ ms	0.98	1.41	1.81	2.17	2.48	2.76	2.98	3.15	3.28
750 ms	0.47	0.69	0.89	1.09	1.27	1.44	1.60	1.75	1.88
1 000 ms	0.27	0.40	0.52	0.64	0.76	0.87	0.97	1.07	1.16
1 250 ms	0.18	0.26	0.34	0.42	0.50	0.57	0.65	0.71	0.78
1 500 ms	0.12	0.18	0.24	0.30	0.35	0.41	0.46	0.51	0.56

$V = 2\ 600$ m/sec

Water Section

$Z =$	40 m	60 m	80 m	100 m	120 m	140 m	160 m	180 m	200 m
$t_1 = 500$ ms	0.72	1.03	1.32	1.57	1.79	1.98	2.12	2.23	2.29
750 ms	0.34	0.50	0.65	0.79	0.92	1.04	1.15	1.25	1.34
1 000 ms	0.20	0.29	0.38	0.47	0.55	0.63	0.70	0.77	0.84
1 250 ms	0.13	0.19	0.25	0.31	0.36	0.42	0.47	0.52	0.56
1 500 ms	0.09	0.13	0.18	0.22	0.26	0,30	0.33	0.37	0.41

$V = 2\ 800$ m/sec

$Z =$	40 m	60 m	80 m	100 m	120 m	140 m	160 m	180 m	200 m
$t_1 = 500$ ms	0.54	0.77	0.98	1.16	1.32	1,45	1.55	1.61	1.64
750 ms	0.25	0.37	0.48	0.58	0,68	0.77	0.85	0.92	0.98
1 000 ms	0.15	0,22	0.29	0.35	0.41	0.47	0.52	0.57	0.62
1 250 ms	0,10	0.14	0.19	0,23	0.27	0.31	0.35	0.38	0,42
1 500 ms	0.07	0.10	0.13	0.16	0.19	0.22	0.25	0.27	0,30

$V = 3\ 000$ m/sec

$Z =$	40 m	60 m	80 m	100 m	120 m	140 m	160 m	180 m	200 m
$t_1 = 500$ ms	0.41	0.58	0.73	0,87	0.99	1.08	1.14	1,18	1.20
750 ms	0.19	0.28	0.36	0.44	0.51	0.58	0.64	0.69	0.73
1 000 ms	0,11	0.16	0.21	0.26	0.31	0.35	0,39	0.43	0.46
1 250 ms	0.07	0.11	0.14	0.17	0.20	0.23	0.26	0.29	0.31
1 500 ms	0.05	0.08	0.10	0.12	0.14	0.17	0,19	0.21	0,23

$$\boxed{X = 500 \text{ m}}$$

$V = 2\,000$ m/sec

Water Section

$Z =$	40 m	60 m	80 m	100 m	120 m	140 m	160 m	180 m	200 m
$t_1 = 500$ ms	2.87	4.18	5.40	6.53	7.56	8.49	9.31	10.03	10.64
750 ms	1.45	2.13	2.77	3.39	3.97	4.53	5.05	5.54	6.01
1 000 ms	0.86	1.27	1.66	2.04	2.41	2.76	3.10	3.42	3.73
1 250 ms	0.56	0.83	1.10	1.35	1.60	1.84	2.07	2.30	2.52
1 500 ms	0.40	0.59	0.78	0.96	1.14	1.31	1.48	1.65	1.81

$V = 2\,200$ m/sec

$Z =$	40 m	60 m	80 m	100 m	120 m	140 m	160 m	180 m	200 m
$t_1 = 500$ ms	2.03	2.94	3.78	4.53	5.21	5.82	6.34	6.76	7.06
750 ms	1.00	1.47	1.92	2.34	2.74	3.11	3.45	3.78	4.08
1 000 ms	0.59	0,87	1,14	1,40	1,65	1,89	2,12	2,34	2,55
1 250 ms	0.39	0,57	0.75	0.93	1.10	1,26	1.42	1.57	1.72
1 500 ms	0,27	0.40	0,53	0.66	0.78	0.90	1,01	1.13	1.24

$V = 2\,400$ m/sec

$Z =$	40 m	60 m	80 m	100 m	120 m	140 m	160 m	180 m	200 m
$t_1 = 500$ ms	1.46	2.11	2.70	3.23	3.70	4.11	4.43	4.68	4.84
750 ms	0.72	1.04	1.35	1.65	1.93	2.19	2.43	2.66	2.86
1 000 ms	0.42	0.62	0.81	0.99	1.17	1.34	1.50	1.65	1.79
1 250 ms	0.27	0.41	0,53	0,66	0.78	0.89	1.00	1.11	1.21
1 500 ms	0.19	0.29	0.38	0,47	0.56	0.64	0.72	0.80	0.88

$V = 2\ 600\ \text{m/sec}$

Water Section

$Z =$	40 m	60 m	80 m	100 m	120 m	140 m	160 m	180 m	200 m
$t_1 = 500$ ms	1.09	1.56	1.98	2.36	2.69	2.97	3.19	3.34	3.41
750 ms	0.53	0.77	1.00	1.21	1.40	1.59	1.76	1.91	2.05
1 000 ms	0.31	0.45	0.59	0.72	0.85	0.97	1.08	1.19	1.29
1 250 ms	0.20	0.29	0.39	0.48	0.56	0.65	0.73	0.80	0.88
1 500 ms	0.14	0.21	0.27	0.34	0.40	0.46	0.52	0.57	0.63

$V = 2\ 800\ \text{m/sec}$

$Z =$	40 m	60 m	80 m	100 m	120 m	140 m	160 m	180 m	200 m
$t_1 = 500$ ms	0.80	1.15	1.48	1.75	1.99	2.18	2.33	2.42	2.46
750 ms	0.39	0.57	0.74	0.90	1.04	1.18	1.30	1.41	1.51
1 000 ms	0.23	0.34	0.44	0.54	0.63	0.72	0.80	0.88	0.95
1 250 ms	0.15	0.22	0.29	0.35	0.42	0.48	0.54	0.59	0.65
1 500 ms	0.10	0.15	0.20	0.25	0.30	0.34	0.38	0.42	0.47

$V = 3\ 000\ \text{m/sec}$

$Z =$	40 m	60 m	80 m	100 m	120 m	140 m	160 m	180 m	200 m
$t_1 = 500$ ms	0.61	0.88	1.11	1.32	1.49	1.63	1.73	1.79	1.80
750 ms	0.30	0.43	0.56	0.68	0.79	0.89	0.98	1.06	1.12
1 000 ms	0.17	0.25	0.33	0.41	0.48	0.54	0.60	0.66	0.71
1 250 ms	0.12	0.17	0.22	0.27	0.32	0.37	0.41	0.45	0.49
1 500 ms	0.08	0.12	0.15	0.19	0.22	0.26	0.29	0.32	0.35

$$X = 600 \text{ m}$$

V = 2 000 m/sec
Water Section

Z =	40 m	60 m	80 m	100 m	120 m	140 m	160 m	180 m	200 m
$t_1 = 500$ ms	3.89	5.68	7.28	8.79	10.17	11.42	12.52	13.46	14.25
750 ms	2.04	2.98	3.88	4.73	5.54	6.32	7.04	7.73	8.38
1 000 ms	1.21	1.78	2.34	2.87	3.39	3.87	4.35	4.81	5.25
1 250 ms	0.80	1.19	1.56	1.92	2.27	2.61	2.95	3.27	3.58
1 500 ms	0.57	0.84	1.11	1.37	1.62	1.87	2.11	2.35	2.58

V = 2 200 m/sec

Z =	40 m	60 m	80 m	100 m	120 m	140 m	160 m	180 m	200 m
$t_1 = 500$ ms	2.76	3.98	5.11	6.14	7.07	7.89	8.56	9.11	9.50
750 ms	1.40	2.05	2.67	3.26	3.81	4.35	4.85	5.30	5.73
1 000 ms	0.82	1.22	1.60	1.97	2.33	2.66	2.99	3.30	3.60
1 250 ms	0.55	0.82	1.07	1.32	1.56	1.80	2.02	2.24	2.45
1 500 ms	0.39	0.58	0.76	0.94	1.11	1.28	1.45	1.61	1.77

V = 2 400 m/sec

Z =	40 m	60 m	80 m	100 m	120 m	140 m	160 m	180 m	200 m
$t_1 = 500$ ms	1.98	2.85	3.65	4.37	5.03	5.57	6.02	6.35	6.54
750 ms	1.00	1.47	1.91	2.32	2.71	3.08	3.41	3.72	4.00
1 000 ms	0.60	0.88	1.15	1.41	1.66	1.90	2.12	2.33	2.54
1 250 ms	0.39	0.58	0.76	0.94	1.11	1.27	1.43	1.58	1.73
1 500 ms	0.28	0.41	0.54	0.66	0.79	0.91	1.02	1.14	1.25

$V = 2\ 600$ m/sec

Water Section

$Z =$	40 m	60 m	80 m	100 m	120 m	140 m	160 m	180 m	200 m
$t_1 = 500$ ms	1.45	2.09	2.68	3.19	3.64	4.01	4.29	4.51	4.63
750 ms	0.73	1.07	1.39	1.69	1.97	2.23	2.47	2.69	2.88
1 000 ms	0.43	0.64	0.84	1.02	1.20	1.38	1.54	1.69	1.84
1 250 ms	0.28	0.42	0.55	0.68	0.80	0.92	1.04	1.14	1.25
1 500 ms	0.20	0.30	0.39	0.48	0.57	0.66	0.74	0.82	0.90

$V = 2\ 800$ m/sec

$Z =$	40 m	60 m	80 m	100 m	120 m	140 m	160 m	180 m	200 m
$t_1 = 500$ ms	1.11	1.59	2.01	2.39	2.71	2.97	3.16	3.28	3.32
750 ms	0.55	0.80	1.04	1.26	1.47	1.66	1.83	1.98	2.11
1 000 ms	0.32	0.48	0.62	0.76	0.89	1.02	1.14	1.25	1.35
1 250 ms	0.21	0.31	0.41	0.51	0.60	0.68	0.77	0.85	0.92
1 500 ms	0.15	0.22	0.29	0.36	0.42	0.49	0.55	0.61	0.67

$V = 3\ 000$ m/sec

$Z =$	40 m	60 m	80 m	100 m	120 m	140 m	160 m	180 m	200 m
$t_1 = 500$ ms	0.85	1.22	1.55	1.83	2.06	2.25	2.37	2.44	2.45
750 ms	0.42	0.61	0.79	0.96	1.11	1.25	1.38	1.49	1.58
1 000 ms	0.25	0.36	0.47	0.58	0.68	0.77	0.86	0.94	1.02
1 250 ms	0.16	0.24	0.31	0.38	0.45	0.52	0.58	0.64	0.70
1 500 ms	0.11	0.17	0.22	0.27	0.32	0.37	0.42	0.46	0.50

$$\boxed{X = 1200 \text{ m}}$$

V = 2 000 m/sec

Water Section

Z =	40 m	60 m	80 m	100 m	120 m	140 m	160 m	180 m	200 m
t_1 = 750 ms	6.01	8.84	11.54	14.03	16.48	18.85	21.06	23.08	25.01
1 000 ms	3.97	5.88	7.74	9.52	11.25	12.92	14.52	16.07	17.55
1 250 ms	2.81	4.16	5.48	6.76	8.00	9.19	10.37	11.51	12.62
1 500 ms	2.07	3.06	4.04	4.99	5.93	6.84	7.72	8.59	9.44
1 750 ms	1.57	2.33	3.08	3.81	4.53	5.23	5.92	6.59	7.25
2 000 ms	1.23	1.83	2.42	3.00	3.57	4.13	4.68	5.21	5.74

V = 2 200 m/sec

Z =	40 m	60 m	80 m	100 m	120 m	140 m	160 m	180 m	200 m
t_1 = 750 ms	4.25	6.26	8.20	9.99	11.69	13.34	14.80	16.22	17.45
1 000 ms	2.83	4.18	5.48	6.73	7.94	9.12	10.23	11.26	12.26
1 250 ms	1.95	2.90	3.82	4.71	5.57	6.41	7.23	8.02	8.78
1 500 ms	1.44	2.13	2.81	3.47	4.11	4.76	5.37	5.97	6.55
1 750 ms	1.09	1.61	2.13	2.64	3.13	3.62	4.09	4.56	5.01
2 000 ms	0.85	1.26	1.67	2.07	2.46	2.85	3.23	3.60	3.96

V = 2 400 m/sec

Z =	40 m	60 m	80 m	100 m	120 m	140 m	160 m	180 m	200 m
t_1 = 750 ms	3.16	4.63	6.02	7.33	8.55	9.69	10.71	11.70	12.54
1 000 ms	2.07	3.04	3.98	4.88	5.74	6.56	7.34	8.08	8.78
1 250 ms	1.42	2.10	2.76	3.39	4.01	4.61	5.19	5.74	6.28
1 500 ms	1.03	1.52	2.00	2.47	2.93	3.38	3.83	4.25	4.66
1 750 ms	0.78	1.15	1.52	1.88	2.23	2.58	2.91	3.24	3.56
2 000 ms	0.61	0.90	1.19	1.47	1.75	2.02	2.29	2.55	2.81

$V = 2\ 600$ m/sec

Water Section

$Z =$	40 m	60 m	80 m	100 m	120 m	140 m	160 m	180 m	200 m
$t_1 = 750$ ms	2.40	3.50	4.53	5.49	6.39	7.22	7.97	8.64	9.23
1 000 ms	1.52	2.24	2.93	3.59	4.22	4.81	5.38	5.91	6.41
1 250 ms	1.05	1.54	2.03	2.49	2.95	3.38	3.80	4.20	4.58
1 500 ms	0.75	1.12	1.47	1.81	2.15	2.47	2.79	3.09	3.39
1 750 ms	0.57	0.84	1.11	1.37	1.63	1.88	2.13	2.36	2.60
2 000 ms	0.43	0.65	0.86	1.07	1.27	1.47	1.66	1.85	2.04

$V = 2\ 800$ m/sec

$Z =$	40 m	60 m	80 m	100 m	120 m	140 m	160 m	180 m	200 m
$t_1 = 750$ ms	1.82	2.65	3.43	4.15	4.82	5.45	6.00	6.49	6.91
1 000 ms	1.15	1.70	2.21	2.69	3.16	3.61	4.02	4.42	4.78
1 250 ms	0.79	1.16	1.52	1.87	2.21	2.53	2.84	3.15	3.43
1 500 ms	0.57	0.84	1.10	1.36	1.61	1.85	2.08	2.31	2.52
1 750 ms	0.43	0.63	0.83	1.03	1.22	1.40	1.59	1.76	1.93
2 000 ms	0.33	0.49	0.65	0.80	0.95	1.10	1.25	1.39	1.52

$V = 3\ 000$ m/sec

$Z =$	40 m	60 m	80 m	100 m	120 m	140 m	160 m	180 m	200 m
$t_1 = 750$ ms	1.42	2.06	2.65	3.20	3.71	4.17	4.58	4.93	5.23
1 000 ms	0.89	1.31	1.71	2.08	2.44	2.76	3.08	3.37	3.64
1 250 ms	0.60	0.89	1.16	1.43	1.68	1.93	2.17	2.39	2.60
1 500 ms	0.43	0.64	0.84	1.04	1.22	1.41	1.58	1.75	1.92
1 750 ms	0.32	0.47	0.63	0.77	0.92	1.06	1.20	1.33	1.46
2 000 ms	0.25	0.37	0.49	0.61	0.72	0.84	0.95	1.05	1.16

CHAPTER 8

SEISMIC EMISSION BY VIBRATORS

J. CASSAND AND M. LAVERGNE

The use of mechanical sources in seismic prospecting is not new, but vibrating sources only recently have become important in seismology, since its utilization is governed by the feasibility of processing long signals in a commercially profitable manner.

Seismic prospecting with vibrators is attractive from various aspects:
— flexibility of implementation, simple stacking, elimination of shotholes;
— no fracturing, hence the possibility of prospecting in a built up area;
— choice of suitable signals which allow only frequencies useful in seismic work to be transmitted in the ground;
— good quality of reflection character in favorable cases.

On the other hand, it displays certain drawbacks with respect to conventional seismic work with explosives which should not be underestimated:
— the surface noise generated by the vibrators is often stronger than that given by buried shots. Hence the source and geophone array are more important than in conventional shooting;
— the lack of seismic quiet periods in industrial areas:
— the inability to employ a gain control important in land recording, because of the long signal utilization, the dynamic range is therefore limited in practice to that of the instrumentation without AGC or programmed gain. Hence the need to sacrifice shallow horizons if we wish to obtain deep horizons;
— rather weak output of vibrating plates, hence the need to sum a large number of vibrations to compensate for the weak seismic energy of each vibration;
— necessity to conduct rapid correlations in the field in order to evaluate the chosen signal and array;
— small dynamic range of the correlators.

These last four points demand a field instrumentation and playback center more complicated than for conventional shooting, hence problems of investment and profitability.

We shall only present here some considerations relative to signal emission by the vibrators and the selection of this signal, the main objective being to obtain as large a penetration as possible while maintaining, after correlation, an equally large resolution.

THEORY OF PULSATING PLATES

A rigid plate which excites the ground with a variable force and communicates movement to it is termed pulsating.

The energy transmitted to the earth appears mainly under three forms:
— compressional wave,
— shear wave,
— surface wave.

Only the energy emitted as a compressional wave is of use in seismic prospecting with vertical component seismometers, the other two wave types constitute the major part of the noise.

With a single vibrator, the energies respectively emitted under the three forms are perfectly defined, and we shall see later how the portion returning as compressional waves is rather weak. On the other hand, with an arrangement of several pulsating plates, the energy emitted as a compressional wave can be reinforced at the expense of the shear and surface waves.

Single pulsating plate. Seismic effectiveness

Consider an elastic, homogeneous and isotropic semi-infinite medium, limited by a plane surface.

The pulsating plate is regarded as circular and rigid and rests without friction on the semi-infinite medium. We finally assume that the distribution of constraints at the contact of the plate and the medium is equal to the static distribution. This is essentially true at seismic frequencies. Let us consider the harmonic components of the force.

Displacement of a pulsating mass-less plate

It can be shown (Bycroft, 1956, p. 336) that the components of displacement u of the pulsating plate exposed to the force $Fe^{j\omega t}$ are written as:

$$u_z = \frac{Fe^{j\omega t}}{2\,\pi\mu r_0} \int_0^\infty (\zeta^2 - 1)^{1/2} \frac{\tau^2 \sin(\zeta r_0)\, J_0(\zeta r)}{F_0(\zeta)}\, d\zeta$$

$$u_r = \frac{Fe^{j\omega t}}{2\,\pi\mu r_0} \int_0^\infty \frac{\zeta \sin(\zeta r_0)\, [2\,(\zeta^2 - 1)^{1/2}\,(\zeta^2 - \tau^2)^{1/2} - 2\,\zeta^2 + \tau^2]}{F_0(\zeta)}\, J_1(\zeta r)\, d\zeta$$

(1)

with:

$$F_0(\zeta) = (2\,\zeta^2 - \tau^2)^2 - 4\,\zeta^2\,(\zeta^2 - 1)^{1/2}\,(\zeta^2 - \tau^2)^{1/2}. \tag{2}$$

In these equations:
u_z and u_r are the vertical and radial components of u;
μ is one of the Lame constants of the earth;
r_0 is the radius of the plate at the base of the vibrator;
τ is the ratio of the compressional wave velocity to that of the shear wave;
J_0 and J_1 are Bessel functions;
ζ is a variable of integration.

The plate being rigid, the vertical displacement of the particles at the contact of the plate is equal to that of the plate u_{pl}. Bycroft has calculated $u_{pl} = u_z$ and obtained:

$$u_{pl} = \frac{F}{\mu r_0} [f_1 (r_0 h, \tau) + jf_2 (r_0 h, \tau)] \tag{3}$$

where the expression $\dfrac{f_1 + jf_2}{\mu r_0}$ depends on the ground and the radius of the plate. It characterizes the admittance of the ground beneath the plate. A description of the functions f_1 and f_2 will be found in Bycroft's Figure 4 (1956) or in a publication more accessible to geophysicists (Fail et al., 1962).

By definition, the "admittance of the ground beneath the plate" will be written as the ratio of the displacement velocity of the plate to the applied force:

$$Y = \frac{v_{pl}}{F}. \tag{4}$$

It is expressed directly as a function of f_1, f_2, μ, and r_0:

$$Y = \frac{\omega}{\mu r_0} (f_2 - jf_1). \tag{5}$$

This is a complex expression, with a real part and an imaginary part increasing at slopes 2 and 1, respectively, with the frequency in the seismic band and types of ground usually encountered.

The admittance is correspondingly stronger as the ground rigidity becomes less and the radius of the pulsating plate reduces.

The displacement of the mass-less pulsating plate under the action of the force F, is written

$$\boxed{u_{pl} = F_s \frac{Y}{j\omega}} \tag{6}$$

For a given force, the plate displacement varies with the admittance. For a given ground type, the displacement is proportional to the excitation force. We have conducted a large number of admittance measurements on numerous ground conditions in the Paris Basin. The theoretical admittances of Bycroft and experimental admittances of two ground types, measured by the 3 ton IFP-CGG vibrator, are shown in Figures 85 and 86. The curves are scaled in frequency.

Displacement of pulsating plate of mass m. Force actually exerted on the ground

The Force F_s impressed on the ground is no longer equal to the force F_e exerted by the vibration generator on the pulsating plate, but may be deduced from it by the relation:

$$F_s = F_e - jm\omega v_{pl} = \frac{F_e}{1 + jm\omega Y} \tag{7}$$

Hence the expression for the displacement of the pulsating plate of mass m:

$$\boxed{u_{pl} = F_e \frac{Y}{j\omega (1 + jm\omega Y)}}. \tag{8}$$

Expression (7) shows that on ground with weak admittance (hard ground) the force F_s exerted by the plate of mass m is practically equal to the force F_e produced by the vibration generator.

For ground types of strong admittance (loose ground), the force F_s is equal to the force F_e only at low frequencies. At high frequencies there is deviation which depends both on the ground type and the pulsating plate.

In Figure 87 we have shown the ratio F_s/F_e for plates of mass 150 and 300 kg, radii of 25 and 50 cm., vibrating on types of ground adequately characterizing those in the Paris Basin. It is seen that even if the amplitude F_e of the excitation force is independent of the vibration frequency, the force F_s transmitted to the ground cuts off above a certain frequency called "frequency of vibrator coupling with the ground."

In order to have good transmission over a wide band of frequencies, Figure 87 shows that it is desirable to take a light plate of large section.

It is readily evident that in order to vibrate at a substantial level, ± 3 tons for instance, it is necessary to keep the pulsating plate pressed to the ground by a static force at least equal to 3 tons. We shall see how a system of elastic linkage allows us to resolve the problem.

It can be shown that the ratio F_s/F_e varies according to the relation:

$$\frac{F_s}{F_e} = \frac{1}{1 + j\omega Y \left(m - \dfrac{R}{\omega^2 - R/M_c} \right)},$$

where: ω :   is the angular frequency of the pulsation;

Y: the ground admittance below the vibrator;

m: the mass of the pulsating plate;

M_c: the mass which loads the pulsating plate to prevent it from separating beyond one g;

R: the rigidity of elastic linkage between the pulsating plate and the loading mass M_c.

At frequencies less than the natural frequency of the mass M_c on the springs R ($\omega^2 \ll R/M_c$) the expression tends toward:

$$\frac{F_s}{F_e} = \frac{1}{1 + j\omega Y (m + M_c)}.$$

The mass M_c is added to the mass of the pulsating plate in the expression which defines the vibrator coupling to the ground, which is detrimental.

On the other hand, at frequencies higher than the natural frequency of the mass M_c ($\omega^2 \gg R/M_c$) the expression tends toward:

$$\frac{F_s}{F_e} = \frac{1}{1 + j\omega Y m}.$$

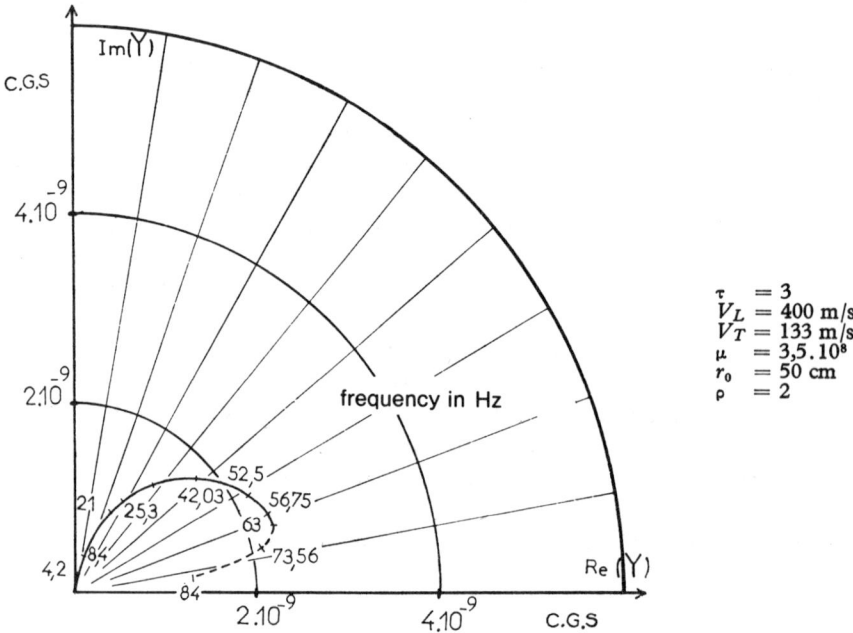

Figure 85.   Theoretical admittances computed by Bycroft (1956).

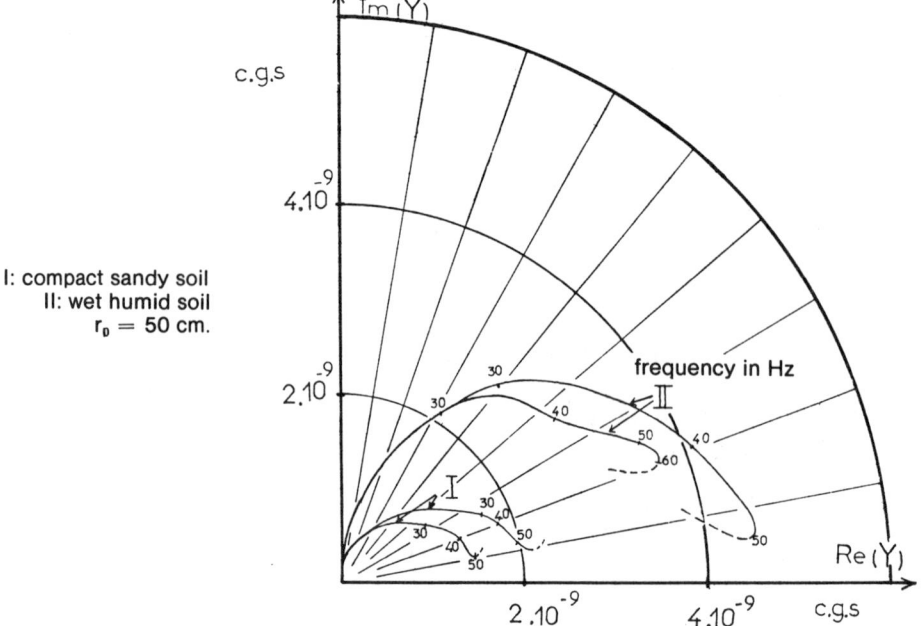

Figure 86.   Admittances measured in the field near Sully-SurLoire.

Only the mass m of the pulsating plate enters into the vibrator coupling with the ground, which is favorable. In practice, it is arranged that the natural frequency $\omega_0 = \sqrt{R/M_c}$ is close to 5 hz. Under these conditions the mass M_c no longer intervenes in the vibrator coupling with the ground in the seismic frequency band, its sole purpose being to exert a static force which prevents the vibrator from detaching at excitations greater than 1 g.

It is feasible in practice to take pulsating plates of mass m sufficiently small so that the ratio F_s/F_e remains close to 1 throughout the entire seismic band.

Because of the elastic linking between the pulsating plate and the load mass M_c, a good coupling is therefore to be had with the ground, even with very heavy vibrators. The load mass M_c can be quite simply that of the transport vehicle.

Note: Control of the force transmitted to the ground.

On certain types of nonuniform ground, in order to have a perfectly reproducible signal essential for summation, it is necessary to have complete control over the force F_s transmitted to the ground. The control is likewise useful when several vibrators are made to work in-phase on nonuniform ground.

To accomplish this control, dynamometers can be placed between the vibrator and the ground, or relation [7] can be employed to recover the excitation force of the vibrator and the inertia force of the plate by appropriate arrangements (pressure gauges, devices to measure displacement, velocity, acceleration, etc.)

In certain cases the control is inadequate, and on soft ground in particular the high frequency components are transmitted very poorly to the ground. It would then be necessary to anchor the pulsating plate by some arrangement of stakes or expansion casings, which would adversely affect the ease of handling the vibrators.

Displacement of ground particles at great distance

The high effectiveness of seismic detection assumes a large movement of the ground particles at great distance, and more precisely, of the radial component of movement for the seismic reflection with vertical component seismometers.

The components of displacement at great distance have been calculated by Miller and Pursey (1954). By neglecting terms in $1/R^2$, they obtain the radial and tangential components in the θ direction (V_L and V_T being the speeds of the compressional and shear waves);

$$\begin{cases} u_R \cong -\dfrac{F_s}{2\pi\mu} \dfrac{\cos\theta}{R} \dfrac{\tau^2 - 2\sin^2\theta}{F_0(\sin\theta)} e^{j\omega(t - R/V_L)}, \\[4mm] u_\theta \cong -\dfrac{F_s}{2\pi\mu} j\tau^3 \dfrac{\sin 2\theta}{R} \dfrac{\sqrt{\tau^2\sin^2\theta - 1}}{F_0(\tau\sin\theta)} e^{j\omega(t - R/V_T)}. \end{cases} \qquad [9]$$

For almost normal incidence, the expression of displacement due to the compressional wave reduces to:

$$u_R \cong -\frac{F_s}{2\pi\rho V_P^2} \frac{1}{R} e^{j\omega(t - R/V_L)}. \qquad [10]$$

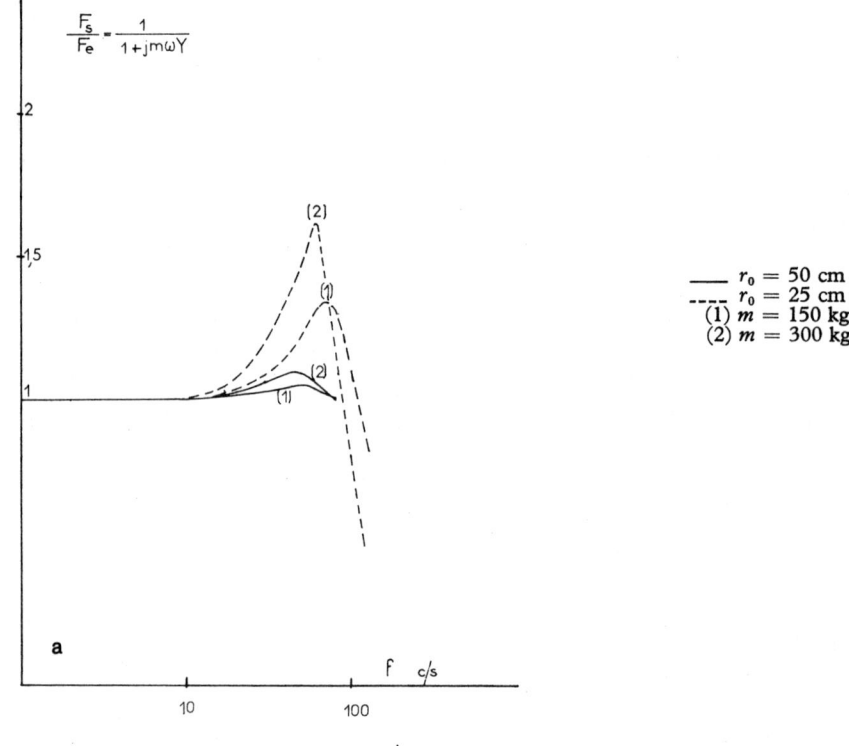

a

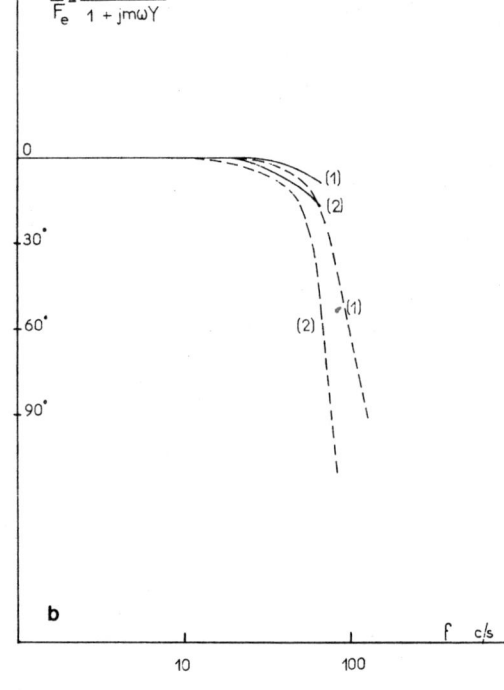

b

Figure 87. Modulus (a) and phase (b) of the ratio F_s/F_e.

Formula [10] shows that the amplitude of the seismic wave diminishes by $1/R$ and that the effectiveness improves as the force becomes greater, as one would expect.

Evidently, this does not take into account absorption and other energy losses, generally more important in loose ground than in rigid ground. Let us recall that the medium was assumed to be elastic, homogeneous, and isotropic in order to expound the theory of pulsating plates, which is far from being representative of the terrain which interests us in exploration. The problem of ground absorption combined with the lack of precise knowledge about the noise prevents us from establishing exactly from first principles what should be expected from the vibrator: a force independent of the frequency, or a force increasing or decreasing with frequency?

For an elastic, homogeneous, isotropic ground without absorption, which is disturbed by noise, giving an equal displacement of particles in the entire band of seismic frequencies, a force of equal amplitude would be required throughout the entire frequency band.

Formula [10] shows that the particle displacement due to the signal is proportional to the force exerted on the ground. But in practice it is known that the ground absorbs the high frequencies more than the low ones, on the other hand the noise generally gives a flatter spectrum when considered in terms of particle velocity rather than particle displacement.

The nature of absorption would therefore lead us to take a force increasing with frequency, while the noise would suggest a force decreasing with frequency. Inasmuch as the two factors enter into play simultaneously, it is not absurd to select an excitation force independent of frequency, especially since we shall see that this solution generally corresponds to optimum use of the vibrator power.

Power distribution in different types of propagation

The total power transmitted into the ground by the pulsating plate and the portion of this power contained respectively in the compressional, shear, and surface waves have been calculated by Miller and Pursey (1954, p. 58) for a semi-infinite elastic, homogeneous, isotropic medium, limited by a surface plane and having a Poisson's ratio of $1/4$ ($\tau = V_L / V_T = \sqrt{3}$).

The manner of calculation is to evaluate the intensities of radiation (power per unit surface) for each type of wave and to integrate over a hemisphere of large radius. The radiation intensity for each wave type is the product of the velocity component and the imaginary conjugate quantity of the component corresponding to the constraint.

The following distributions are obtained:

Compressional wave: $\quad W_L = 0.333 \, \dfrac{\omega^2 \, F_s^2}{4\pi\rho \, V_P^3},$

Shear wave: $\qquad W_T = 1.246 \, \dfrac{\omega^2 \, F_s^2}{4\pi\rho \, V_P^3},$

Surface wave: $$W_S = 3.257 \frac{\omega^2 F_s^2}{4\pi\rho\, V_P^3},$$

Total power: $$W_0 = 4.836 \frac{\omega^2 F_s^2}{4\pi\rho\, V_P^3}.$$

The total power can otherwise be calculated directly from the formula:

$$W_0 = \frac{1}{2} F_s^2 R_e (Y).$$

With a single pulsating plate it is seen that less than 7 percent of the total power is transmitted in the form of a compressional wave, whereas close to 68 percent is in the form of a surface wave.

In order to increase the proportion of energy emitted in the form of a compressional wave at the expense of the surface noise, we can shift the pulsating plate and sum the vibrations, or employ several plates vibrating in-phase at suitable distances.

Several pulsating plates. Improvement in seismic effectiveness

To calculate the displacement at large distance given by several plates vibrating in-phase, it suffices to add vectorially the elemental shifts due to each of the plates.

Miller and Pursey (1955, p. 62) have performed the calculation and have given the power distribution for the three types of waves, as well as the total power transmitted into the ground by several vibrators arranged in a circle of small radius within the distance of observation. The results are quite easily extended to several vibrators placed in line, and we have portrayed in Figure 88 the radiation power for compressional, shear, and surface waves as a function of the spacing d between pulsating plates for:

two vibrators in line,
three vibrators in line,
three vibrators in a triangle.

These results are valid for a Poisson's ratio of 1/4:

$$(V_L/V_T = \tau = \sqrt{3}).$$

It is seen that the power emitted by the compressional wave diminishes rather little with vibrator separation. By contrast, the power emitted by the surface wave decreases quite perceptibly and passes through a minimum for $k_1 d$ close to 2.2. We therefore have the best signal to surface noise ratio for a distance between vibrators $d = 0.35$ times the wavelength of the compressional wave.

In the case of the three vibrators arranged in a triangle, the proportion emitted in the form of a compressional wave reaches 30 percent of the total power when they are located at this optimum distance, but the total power emitted is only 1.5 times the power of a single vibrator (Figure 88).

To eliminate 20 cycle surface noise, the optimum distance between vibrators is about 14 m for a compressional velocity of 800 m/sec (surface velocity in the order of 400 m/sec) and about 28 m for a compressional velocity of 1600 m/sec (surface velocity in the order of 800 m/sec).

These results are of course theoretical since they apply to elastic, homogeneous, isotropic ground, limited by a plane surface, and whose Poisson's ratio is 1/4.

In practice, the ground is rather different from these conditions and it will always be necessary to establish experimentally the optimum distances between vibrators in order to eliminate surface noise.

Note: No contradiction need be seen in the fact that with n vibrators, very closely spaced, the radiation power is theoretically n^2 times that of one vibrator. This is due to the fact that the displacement under each individual vibrator is modified by the presence of the others, in amounts governed by the mutual admittances.

The mutual admittances of two identical plates i and j, located on the ground and vibrating in-phase with the same force, is defined by the ratio of the jth plate velocity to the force applied to the ith plate.

$$Yji = Yij = \frac{v_j}{F_i}.$$

Miller and Pursey (1954, p. 63) have calculated the mutual admittances of two plates in the case where the Poisson coefficient is 1/4 and where the radii are small in relation to the distance of the plates.

The mutual admittances depend on the elastic properties of the medium and the distance of the plates. In Figure 89 we have shown the mutual admittance of two plates vibrating on an elastic, homogeneous, and isotropic ground, limited by a plane surface whose Poisson coefficient is 1/4.

When d is close to zero, we recover the admittance of a single plate. Likewise, the radiation energy of a single pulsating plate can be written:

$$W_0 = \frac{1}{2} F_s^2 R_e(Y),$$

the radiation energy of n plates vibrating in-phase is written:

$$W = \frac{1}{2} F_s^2 \sum_{i,j} R_e(Yij).$$

For three identical vibrators, we have by example:

$$W = \frac{1}{2} F_s^2 [3 R_e(Y) + 6 R_e(Y_{\text{mut}})].$$

The total radiation power is therefore immediately inferred from the mutual admittances (Figures 88 and 89).

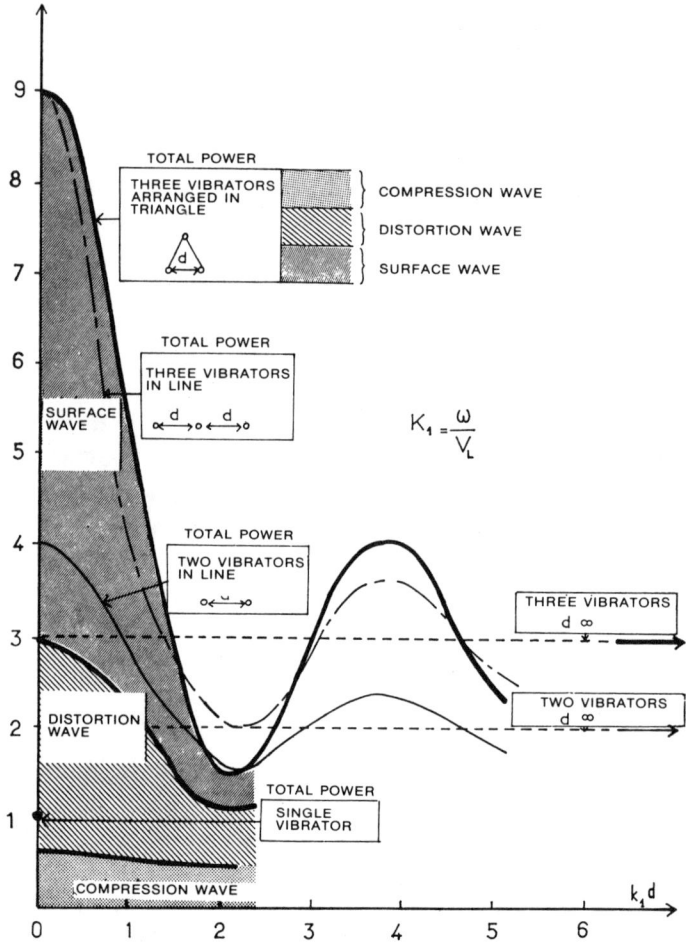

Figure 88.   Radiation power of several vibrators in-phase.

SELECTION OF SEISMIC SIGNAL

The purpose of seismic exploration is to determine with the maximum precision possible the impulse response of the subsurfaces, which will permit reconstruction of the geology from the seismic wave velocity distribution, that is the depth distribution of reflection coefficients when working with normal incidence as in the case of reflection seismology.

In conventional seismic work, this result is attained quite simply by sending into the ground an impulse $s(t)$, generated by an explosion or a weight drop.

The recorded seismogram is the result of convolving the emitted signal $s(t)$ with the impulse response $k(t)$ of the subsurface, to which noise $\xi(t)$ is added, i.e.:

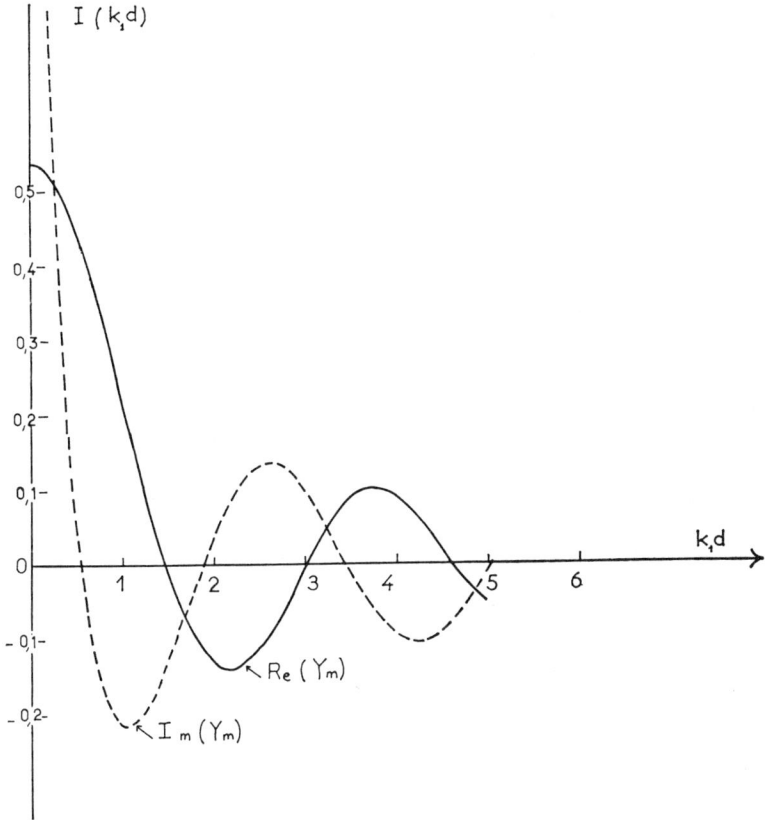

Figure 89.   Mutual admittance of two circular plates.

$$Y_{\text{Mutual}} = -\frac{2\pi j f^2 \tau^4}{\rho V_C^3} I(k_1 d).$$

$$r_0 \ll d$$
$$\tau = \sqrt{3}$$

$$y(t) = \int_{-\infty}^{+\infty} k(\mathcal{C}) s(t - \mathcal{C}) d\mathcal{C} + \xi(t), \qquad (1)$$

This can be written conventionally:

$$y(t) = k(t) * s(t) + \xi(t). \qquad (2)$$

The signal received $y(t)$ will approach the impulse response $k(t)$ as $s(t)$ tends to approximate a unit impulse.

In conventional shooting, if the signal $s(t)$ represents the particle velocity, it has the form more or less of a Ricker impulse. The impulse shape can hardly be modified, but the signal-to-noise ratio is improved by increasing the charge.

In the case of the weight drop, as for the explosive, the shape of the signal cannot be altered and its amplitude can be weak. Improvement in the signal-to-noise ratio is achieved by summing the raw recordings.

In seismic exploration by vibrating, the power emitted is extremely weak, but on the other hand a long signal can be introduced to transmit an appreciable quantity of energy into the ground.

The basic recording obtained will also be represented by formula (1). But here, in contrast to conventional methods, the instantaneous amplitudes are extremely weak, hence the necessity of increasing in the first place the signal-to-noise ratio during recording by summation of the raw records. This filtering of random noise is furthermore combined with wavenumber filtering of the coherent noise by shifting the vibrator.

We shall see later that the method of correlation (and it is this which facilitates use of vibrators) enables us to improve the S/N ratio and thus detect very weak signals buried in the noise. On the other hand, the signal employed does not allow us to obtain the impulse response directly. It is necessary to process the ground recordings in order to compress the long signal into a short impulse. In essence, the emitted signal must be sought in the signal received, to evaluate the degree of correlation between these two functions $s(t)$ and $y(t)$ during their variation in time.

This can be accomplished by crosscorrelation of the signal emitted and the signal received. As a matter of fact, it is known that the impulse response of a system can be reconstructed by performing the crosscorrelation of the input and output signals if the autocorrelation of the input signal is a Dirac impulse. It is this latter condition which determines the choice of signal; the autocorrelation should therefore be as close as possible to a unit impulse.

We are going to demonstrate the foregoing in concise form and deduce results from it concerning the choice of signal giving, at the same time good resolution and a perceptible improvement in the signal-to-noise ratio.

Reconstruction of the impulse response

Autocorrelation of a signal $s(t)$ is defined as the integral:

$$\rho(t) = \int_{-\infty}^{+\infty} s(\tau)\, s(t + \tau)\, d\tau. \tag{3}$$

We note that:

$$\rho(0) = \int_{-\infty}^{+\infty} [s(\tau)]^2\, d\tau \tag{4}$$

represents the total energy of the signal.

It can be demonstrated that $\rho(t)$ is an even function of maximum amplitude equal to $\rho(0)$.

Symbolically we write down:

$$\rho(t) = s(t) * s(-t).\tag{5}$$

Likewise, crosscorrelation of the function $y(t)$ with the function $s(t)$ is defined as the integral:

$$r(t) = \int_{-\infty}^{+\infty} y(\tau) s(t + \tau) d\tau.\tag{6}$$

that is:

$$r(t) = y(t) * s(-t).\tag{7}$$

It is this last operation which is carried out at the processing center to reconstruct the impulse response. In fact, according to equation (2) we have for (7):

$$r(t) = [k(t) * s(t) + \xi(t)] * s(-t)$$

where:

$$r(t) = k(t) * s(t) * s(-t) + \xi(t) * s(t)$$

that is:

$$r(t) = k(t) * \rho(t) + \xi(t) * s(-t).\tag{8}$$

Therefore if $\rho(t)$ approximates a unit Dirac impulse, the impulse response of the earth is approximated by crosscorrelating the emitted and received signals.

Crosscorrelation of the field seismogram with the emitted signal amounts to convolving the impulse seismogram $k(t)$ with the autocorrelation $\rho(t)$ of the emitted signal and correlating the noise $\xi(t)$ with the signal $s(t)$. The final seismogram, resulting from the crosscorrelation, will then be identical to what would have been obtained with a source generating the signal $\rho(t)$. Furthermore, the noise has been filtered by the signal $s(t)$.

Theoretical choice of signal. Examples of long signals presenting a favorable autocorrelation with vibrator seismic work

From formula (8) we can infer the qualities required of the signal autocorrelation and hence of the signal itself.

a) In order to obtain good resolution, the signal autocorrelation should be narrow.

b) To have a good S/N ratio, $\rho(t)$ should display a strong central amplitude $\rho(0)$.

The first condition prescribes a signal covering a wide band of frequencies.

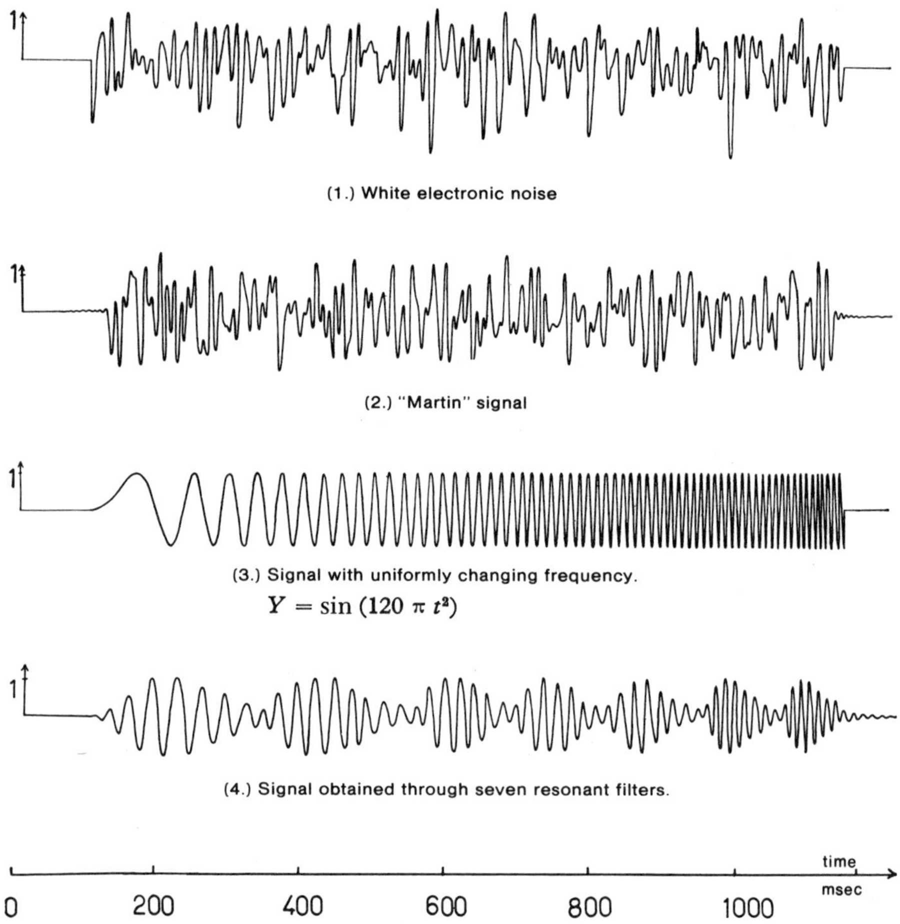

(1.) White electronic noise

(2.) "Martin" signal

(3.) Signal with uniformly changing frequency.

$$Y = \sin (120 \, \pi \, t^2)$$

(4.) Signal obtained through seven resonant filters.

						time
0	200	400	600	800	1000	msec

Figure 90a.   Examples of long signals.

A narrow impulse displays a spectrum rich in frequencies. But the spectrum of the signal autocorrelation is proportional to the square of the signal spectrum modulus. Hence it is necessary to have a signal displaying a wide spectrum.

To fit with the second condition, in order to have a good S/N ratio, the middle amplitude of the autocorrelation must be strong, i.e., the total signal energy must be as high as possible (relation 4). The vibrator being able to emit only rather weak power, a long signal must therefore be employed.

We note that the noise is always filtered by the signal for the purpose of improving the S/N ratio. A judicious choice of signal spectrum should therefore facilitate removal of certain frequencies hampered by noise, in particular the high frequencies.

There are certain signals of several seconds duration whose spectrum coincides

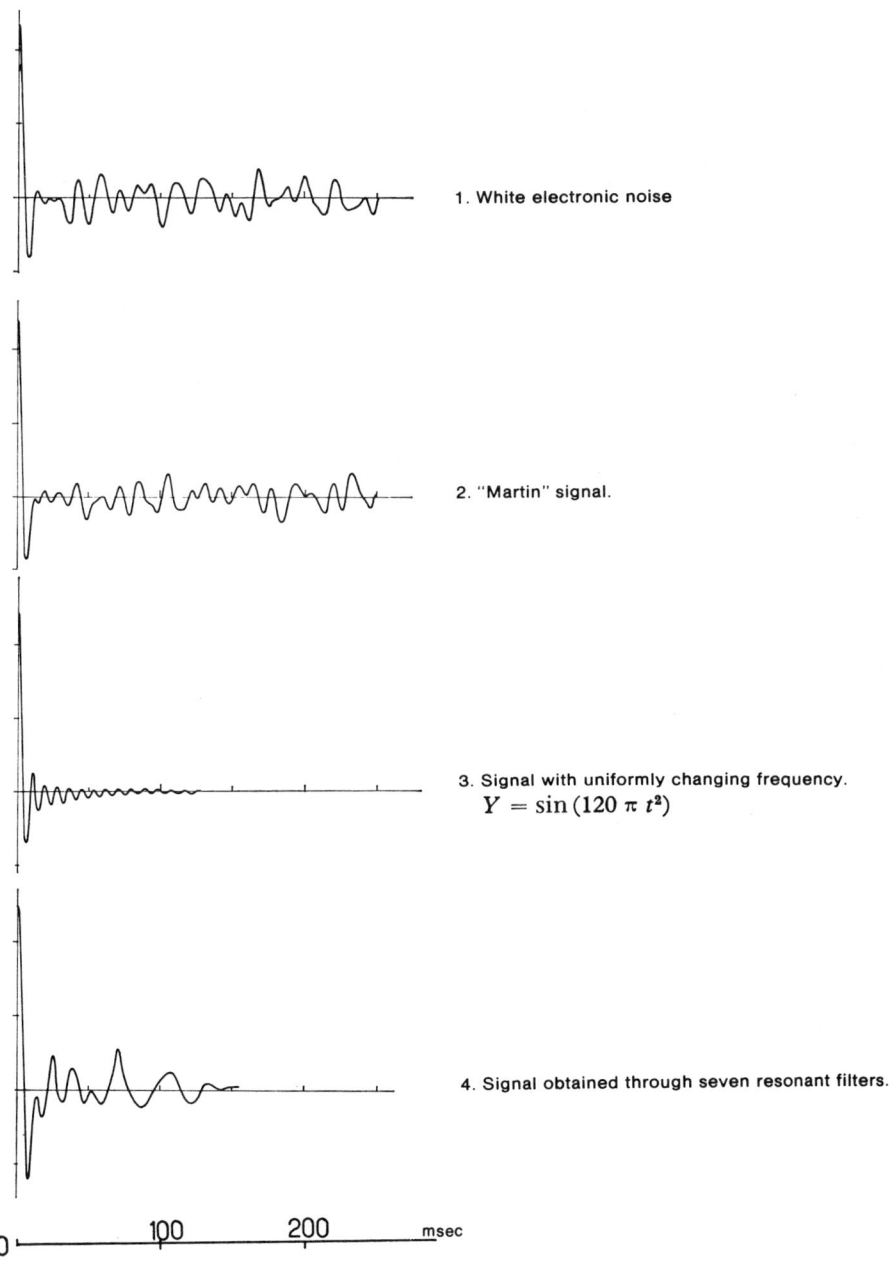

1. White electronic noise

2. "Martin" signal.

3. Signal with uniformly changing frequency.
   $$Y = \sin(120\,\pi\,t^2)$$

4. Signal obtained through seven resonant filters.

0   100   200   msec

Figure 90b.   Autocorrelations of signals in Figure 90a.

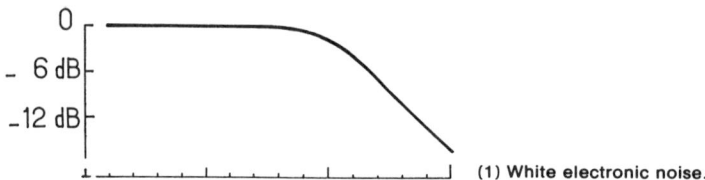

(1) White electronic noise.

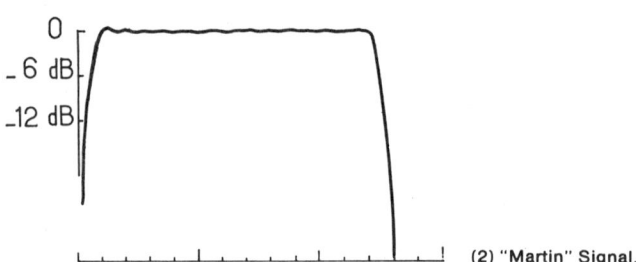

(2) "Martin" Signal.

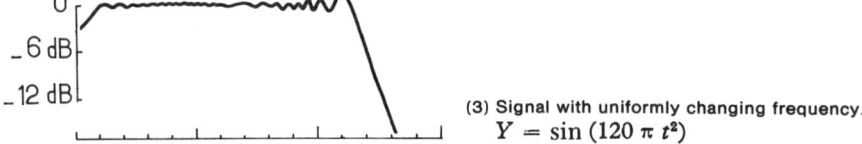

(3) Signal with uniformly changing frequency.
$$Y = \sin(120\,\pi\,t^2)$$

(4) Signal obtained through seven resonant filters.

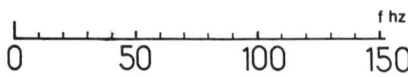

f hz

0   50   100   150

Figure 90c.   Spectra of signals in Figure 90a.

with the useful seismic frequencies and whose autocorrelation is of high amplitude and short duration (30 to 50 msec).

As an example, we have displayed four possible signals in Figures 90a, 90b and 90c with their autocorrelations and spectra.

1. Experimental electronic white noise.

2. Martin signal: computed white noise, formed by a series of unit impulses with randomly distributed sign, spaced 2 msec apart and filtered with a computed passband whose spectrum is flat between 10 and 120 hz (Martin, 1959).

3. Signal with uniformly changing frequency $y = \sin 120 \, \pi \, t^2$. General form $y = \sin (at^2 + bt + c)$.

4. Signal obtained by tuned filters and delay lines.

The properties of the signals considered are essentially equivalent. However, the signal with uniformly changing frequency displays a more correct autocorrelation; its characteristics are well known and it is easily handled. These are the only points in its favor. On the other hand, the signal obtained by tuned filters and delay lines facilitates correlation by means of a simple delay filter.

Effect of absorption in the earth. Its influence on resolution. Compensation for effect.

Absorption in the earth limits resolution of the final seismogram as in conventional seismic work.

Let us assume, for simplicity, that the earth acts as a linear filter $F(\omega)$. The $F(\omega)$ filtering is going to be encountered after correlation on the final seismogram which will not be the convolution of the impulse seismogram $k(t)$ with the autocorrelation $\rho(t)$, but with the autocorrelation $\rho(t)$ filtered by the $F(\omega)$ filter representing the absorption. This, of course, has the effect of appreciably stretching the signal $\rho(t)$ and therefore of diminishing the resolving power.

In Figure 91 we have displayed the theoretical broadening of the signal $\rho(t)$ for a more or less absorbent earth. In practice, it is not unusual that the signal broadening reaches 60 msec between the two autocorrelation peaks. We run into the same difficulty here as in conventional seismic work with explosives.

We can of course compensate for the absorption of the high frequencies and improve the resolution by reinforcing the proportion of high frequencies in the emitted signal, for example by concentrating more on the high frequencies than on the lows in a signal with nonuniformly variable frequency. We can likewise perform conventional operations of deconvolution on the final seismogram. We can also plan a very slow gain program favoring the high frequencies during field recording. It will then be necessary to transmit frequencies increasing with time.

Nevertheless, the differential absorption of frequencies in the earth always limits resolution as in conventional seismic work. Inasmuch as the emitted signal is limited in the time domain, it will be futile to attempt to improve resolution if the earth attenuates the high frequencies too strongly. The quest for an often illusory improvement in resolution does not proceed without deteriorating the S/N ratio.

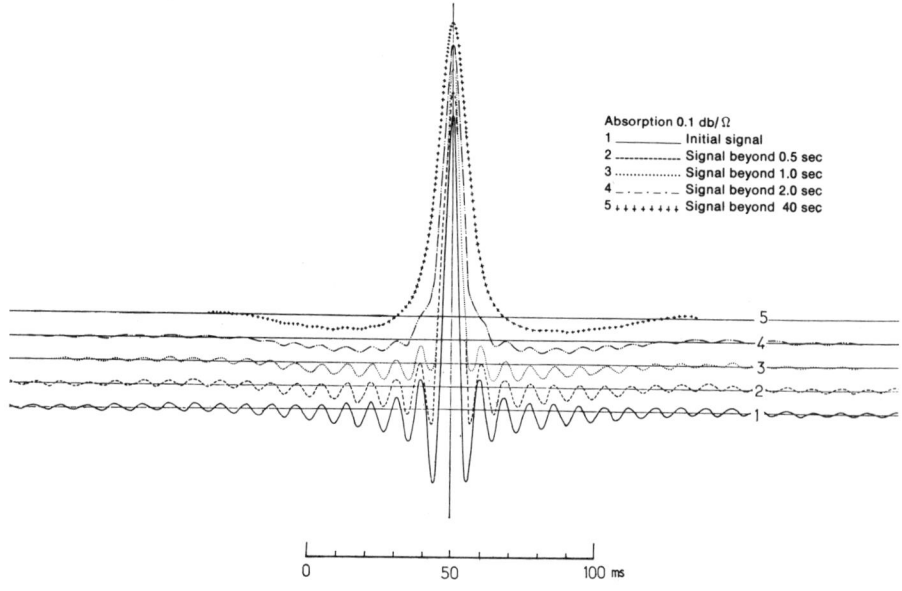

Figure 91.   Influence of absorption on the autocorrelation.

Another factor likely to limit resolution is the difficulty, previously indicated in the paragraph on pulsating plates, in sending high frequencies into soft ground, because of the poor vibrator coupling.

Effect of signal distortion

As we shall see, the crosscorrelation is generally performed with the vibrator pilot signal. But vibration generators sometimes have a certain distortion and the seismic signal is not strictly identical to the pilot signal.

This phenomenon would be troublesome if it should disturb the restoration of the final seismogram. Figure 92 shows that such is not the case: we have displayed the autocorrelation of the signal $s(t) = \sin 120 \pi t^2$ and its crosscorrelation with the signal $s^*(t)$ obtained by applying to $s(t)$ a harmonic distortion ratio of 35 percent.

It is seen that autocorrelation and crosscorrelation with the distorted signal have identical resolving power. The amplitude of the crosscorrelation is slightly weaker than that of the autocorrelation.

Distortion of the vibration generators is generally limited to 15 or 20 percent in the most unfavorable cases. We can therefore infer that this phenomenon has a rather small effect on the final seismogram, particularly as the harmonics are partly eliminated by selective absorption of the high frequencies during transmission.

Correlation through a simplified form of emitted signal

Simplification of the correlation operation may entail crosscorrelation not with the emitted signal $s(t)$ but with a modified signal which characterizes $s(t)$ but occurs in a more tractable form (discrete samples, spikes, etc.)

Choice of reference signal employed for correlation. Pilot signal or signal actually transmitted into the ground

Generally in seismic prospecting the signal conveyed by the geophones represents the velocity of the ground. It would then seem logical to utilize a similar dimension as a reference signal for correlation, that is, a velocity.

We can imagine, for example, taking the velocity of the pulsating plate for the reference signal, that is, the velocity imparted to the ground by the vibrator. A velocity pick-up located on the plate or in the immediate vicinity would furnish the signal.

Now the ground behaves approximately like a spring. If the vibrator is subjected to a force, the plate displacement will be proportional to this force, but the velocity will be proportional to the derivative of this force with respect to time. The autocorrelation of such a signal is often not very suitable for correlation.

The same applies to the signal delivered by a geophone in the vicinity of the vibrator. Although attenuated, this signal favors the high frequencies but even now displays amplitude variations due to the fact that the signal has been filtered by the earth in the immediate vicinity of the plate (reflections from shallow beds, surface waves, local diffractions, absorption of the high frequencies, etc.). It is therefore difficult to employ this signal as a reference.

One solution would consist of using a displacement indicator located on the plate. The possibility also remains of performing the correlation with the acceleration of the mass or with the electrical pilot signal (signal of constant amplitude); this last solution is the simplest since there is no need to carry out additional recording, and moreover, the results obtained in practice are satisfactory.

In Figure 93 we have displayed the autocorrelations of some usable reference signals.
1. ground velocity at 1 meter from the plate;
2. acceleration of the plate;
3. acceleration of the mass;
4. pilot signal.

The signals transmitted by the geophone and by the plate-accelerometer are distorted at low frequencies. Their autocorrelations are very poor, hence the inability to use them as a reference. We observed that the distortion of the signal disappears in the course of propagation. The strong amplitude of the high frequencies offers the benefit of compensating somewhat for the absorption. The autocorrelation of the mass acceleration signal is almost as good as the autocorrelation of the pilot signal.

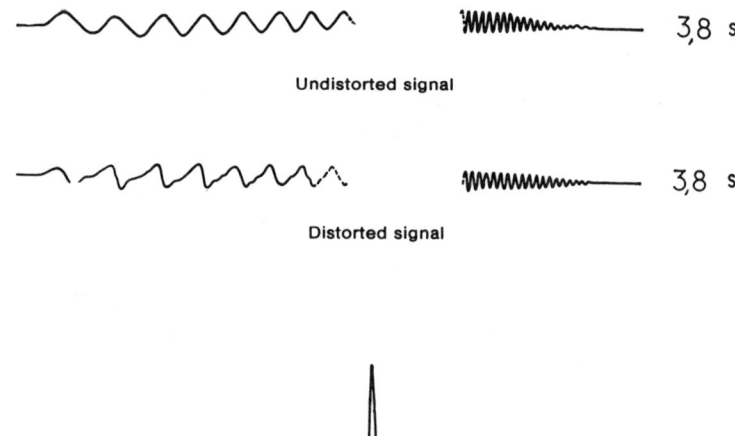

Undistorted signal

Distorted signal

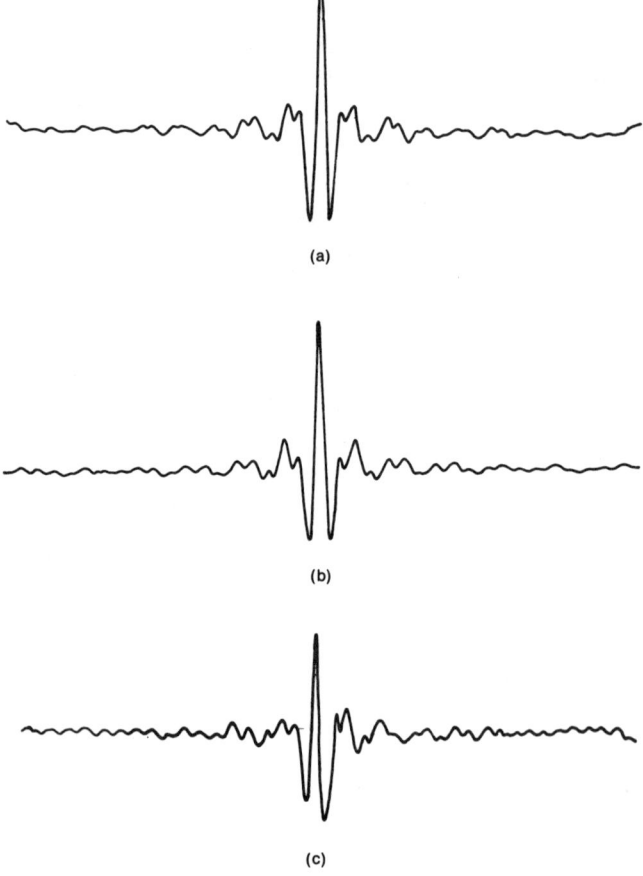

Figure 92. Autocorrelations of a nondistorted signal (a) and a signal distorted 35 percent, (b) and the crosscorrelation of the two signals (c).

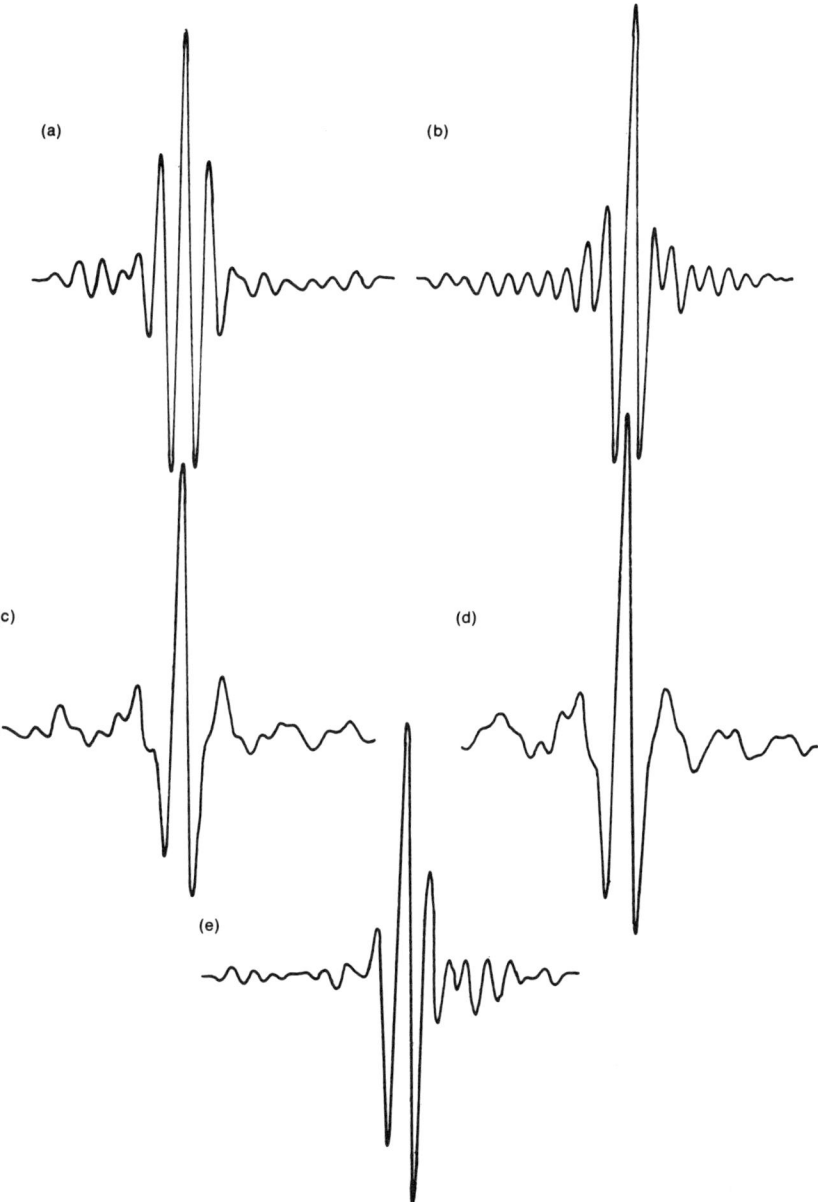

Figure 93.   Autocorrelations of the geophone velocity one meter from the plate (a), of the plate acceleration (b), of the acceleration of the mass (c), and of the input signal (d); and the crosscorrelation of the input signal with the geophone velocity trace (e).

The crosscorrelation of the pilot signal with the ground velocity one meter from the plate is rather poor. However, since the high frequencies are filtered in the ground, the signal will display a flatter spectrum and consequently a better correlation, for a marker at average depth.

A force-controlled vibrator is adapted to a depth of penetration for which the absorption has equalized, in some way, the spectrum of the transmitted signal. The important thing is to choose a reference of constant amplitude. The pilot signal is therefore suitable.

BIBLIOGRAPHY

Agard, J., and Grau, G., 1961, Statistical study of seismograms: Geophys. Prosp., v. 19, p. 503-525.

Allen, W. B., and Westerfield, E. C., 1964, Digital compressed-time correlators and matched filters for active sonar: J. Acoust. Soc. of America, v. 36, p. 121-139.

Angot, A., 1957, Compliments de mathematiques, 3rd edition: Editions de la Revue d'optique.

Anstey, N. A., 1963, Vibroseis gentle massage obtains structural data safely, economically: Oil and Gas J., v. 61, p. 110-118.

Arsac, J., 1961, Fourier transformation and distribution theory, Paris, Dunod.

Backus, M. M., 1959, Water reverberations; their nature and elimination: Geophysics, v. 24, p. 233-261.

Barbier, M.G.Y., Haskett, R. W., Chavy, J., and Perez, 1964, Le Procede Vibroseis, Mise en oeuvre et parametres d'enregistrement: Section Geol. Geophys. Bull. AFTP, p. 389-402.

Bois, P., and Hemon, Ch., 1964, Condition to be fulfilled by a function in order to represent an impulse seismogram with multiple reflections, for any surface reflection coefficient: Annales de Geophysique, v. 20, p. 509-511.

Bois, P., Grau, G., Hemon, Ch., and La Porte, M., 1962, Automatic calculation of synthetic seismograms of plane waves with normal incidence: Rev. Inst. France du Petrole, v. 27, p. 491-522.

Boll, 1957, Tables numeriques universelles, 2nd edition: Paris, Dunod.

Bracewell, R., 1965, The Fourier transform and its applications: New York, McGraw-Hill Book Co., Inc.

Brillouin, L., 1962, Science and theory of information: Academic Press.

Burg, K. E., Ewing, M., Press, P., and Stulken, E. S., 1951, A seismic wave guide phenomenon: Geophysics, v. 16, p. 594-611.

Bycroft, G. N., 1956, Forced vibrations of a rigid circular plate on a semi-infinite elastic space and on an elastic stratum: Phil. Trans. Roy. Soc., no. 248, p. 327-368.

Chauveau, J., 1963, Filtering compensator in the propagation of a seismic signal: Annales de Geophysique, v. 19, no. 1, p. 21-30.

Continental Oil Company, French Patent No. 1.276.377, 1960.

Crawford, N. J., Doty, W. E. N., and Lee, M. R., 1960, Continuous signal seismograph: Geophysics, v. 25, p. 95-105.

Dobrin, M. B., Ingalls, A. L., and Long, J. A., 1965, Velocity and frequency filtering of seismic data using laser light: Geophysics, v. 30, p. 1166-1178.

Domenico, S. N., 1963, Phase-distortionless filtering: Geophysics, v. 30, p. 35-50.

Ellis, L. G., and Winterhalter, A. C., 1956, Unusual reflection events in offshore seismic work: Geophysics, v. 21, p. 755-765.

Embree, P., Burg, J. P., and Backus, M. M., 1963, Wide-band velocity filtering — the pic-slice process: Geophysics, v. 28 p. 948-974.

Fail, J. P., Grau, G., and Layotte, P. C., 1964, Amelioration du rapport signal/bruit a l'aide du filtrage in eventail — Etude d'un cas concret: Geophys. Prosp., v. 12, p. 258-282.

Fail, J. P., and Grau, G., 1963, Les filtres en eventail: Geophys. Prosp., v. 11, p. 131-163.

Fail, J. P., Grau, G., and Lavergne, M., 1962, Seismograph coupling with the ground: Geophys., Prosp. v. 10, p. 128-141.

Finn, R. S., and Heap, W. O., 1962, How vibratory systems are performing: World Oil, v. 154, p. 108-111.

Goupillaud, P. L., and Lee, M. R., 1963, Some theoretical aspects of the vibroseis system and their practical applications: Paper presented at the 33rd Annual International SEG Meeting, New Orleans.

Grau, G., and Herman, Ch., Commercial deconvolution (in preparation).

Jackson, P., 1965, Analysis of variable-density seismograms by means of optical diffraction: Geophysics, v. 30, p. 5-23.

Jennison, R. C., 1961, Fourier transforms and convolution for the experimentalist: London, Pergamon.

Jones, H. S., Morrison, J. A., Sarrafian, G. P., and Speiker, L. J., 1955, Magnetic delay line filtering techniques: Geophysics, v. 20, p. 745-765.

Klauder, J. R., Price, A. C., Darlington, S., and Albersheim, W. J., 1960, The theory of design of chirp radars: Bell System Tech. J., v. 39, p. 745-808.

Kunetz, G., 1964, Generalization of dereverberation operators with any number of reflectors: Geophys. Prosp., v. 12, p. 283-289.

——— 1963, Some examples of seismic record analysis: Geophys. Prosp., v. 11, p. 409-422.

——— 1961, Analysis test of seismic traces: Geophys. Prosp., v. 9, p. 317-341.

Kunetz, G., and d'Erceville, Y., 1962, Certain properties of a plane compressional acoustic wave in a stratified medium: Annales de Geophysique, v. 18, p. 351-359.

Labin, E., 1949, Operational calculus: Paris, Masson.

Lee, M. R., and Crawford, J. M., 1963, Vebroseis application becoming worldwide: World Petroleum, v. 34, p. 40-41.

Lindsey, J. P., 1960, Elimination of seismic ghost reflections by means of a linear filter: Geophysics, v. 25, p. 130-140.

Maas, H. W., 1964, Das Vibroseis System als optimales Laufzeit-Verfahren: Erdöl und Kohl, v. 17, p. 699-707.

Martin, M. A., 1959, Frequency domain applications to data processing: IRE transactions on space electronics telemetry, no. 1.

Miller, G. F., and Pursey, H., 1954, The field and radiation impedance of mechanical radiators on the free surface of semi-infinite isotropic solid: Proc. Roy. Soc., no. 223, p. 521-541.

Miller, G. F., and Pursey, H., 1955, On the partition of energy between elastic waves in a semi-infinite solid: Proc. Roy. Soc., no. 233, p. 55-69.

Papoulis, A., 1962, The Fourier integral and its applications: New York, McGraw-Hill Book Co., Inc.

Rice, R. B., 1962, Inverse convolution filters: Geophysics, v. 27, p. 4-18.

Rybner, J., 1946, Fourieranalyse af frekvensmodulerede svingninger med savtakvariation af ojebliksfrekvensen (kipmodulation): Akad. für Tekniske Videnskaber Lydteknisk Laboratorium, no. 2.

Sapirovskij, N. I., and Gadziev, P. M., 1962, Geophysical prospecting at sea: Bakou, Azernesr.

Schneider, W. A., Larner, K. L., Burg, J. P., and Backus, M. M., 1964, A new data-processing technique for the elimination of ghost arrivals on reflection seismograms: Geophysics, v. 29, p. 783-805.

Schwartz, L., 1961, Mathematical methods for the physical sciences: Paris, Hermann.

Sutton, G. M., Berckhemer, H., and Nafe, J. E., 1957, Physical analysis of deep sea sediments: Geophysics, v. 22, p. 779-812.

Valiron, G., 1948, Theorie des functions, 2nd edition: Paris, Masson.

Van der Lugt, A., Signal detection by complex spatial filtering: IEEE Trans. Info. Theory, v. 11, no. 2, p. 139-145.

Werth, G. C., Liu, D. T., and Trorey, A. W., 1959, Offshore singing, field experiments and theoretical interpretation: Geophysics, v. 24, p. 220-232.

Winkell, 1960, Vues nouvelles sur le monde des sons: Paris, Dunod.

Wuenschel, P. C., 1960, Seismogram synthesis including multiples and transmission coefficients: Geophysics, v. 25, p. 106-129.